Bringing Power to the People

The History of IBEW Local #77

Ellie Belew

International Brotherhood of Electrical Workers Local #77
Seattle, Washington

 1

International Brotherhood of Electrical Workers #77
P.O. Box 68728
Seattle, Washington 98168
© 2013 by IBEW #77
All rights reserved. Published 2013.

Printed in the United States of America

ISBN 978-0-9892859-0-2

Printed at Capitol City Press
Tumwater, Washington 98512
http://www.capitolcitypress.com/contact-us/

Table of Contents

Recent Accomplishments

Appendices

PSP&L safety warning, site and date unknown.
University of Washington Special Collections, UW35553

Foreword

As Ellie Belew's history of IBEW Local #77 shows, the ability of workers to receive a living wage, protect their job from arbitrary bosses, maintain a safe working environment, gain health and retirement coverage, and be of service to the broader community does not come from the employer. It comes from the workers themselves.

It is the willingness of workers to take collective and strategic action beyond the worksite that makes all these things possible. Without this willingness, employers are quite willing to work us to death for free.

IBEW Local 77 is a democratic organization. Generations of workers have built it so that workers, if they choose to, can see themselves as a group, analyze the position they are in with regard to their employer and their community, debate strategy and tactics, and take disciplined action based on solidarity.

Because IBEW workers have chosen to act collectively through their union, they have helped shape Washington State, its heritage and institutions. Without knowing IBEW #77's history, one cannot explain the Seattle General Strike of 1919, the innovative industrial union model, the tremendous electrification of rural Washington in the 1920s and 1930s, the BPA transmission grid that facilitated regional development, the establishment of public utility districts after World War II, or the availability of firm, reliable electrical energy for Washington's manufacturing base, its aluminum industry, and its modern internet technology sector.

Who climbed all those poles, erected all those towers, strung all that wire, keep track of all that revenue, answered all those worried phone calls, and risked their lives in the midst of storms to restore power? It wasn't PUD commissioners or executives from private electric monopolies. It was IBEW #77.

I was going to say this book is a gift to us from IBEW #77. But it is not a gift. It is a necessity. As global finance seeks to change the very character of our communities and work life, only by knowing how we got this far will we figure out how to survive in the future.

Dan Leahy
Founder and former Director
The Evergreen State College Labor Education and Research Center

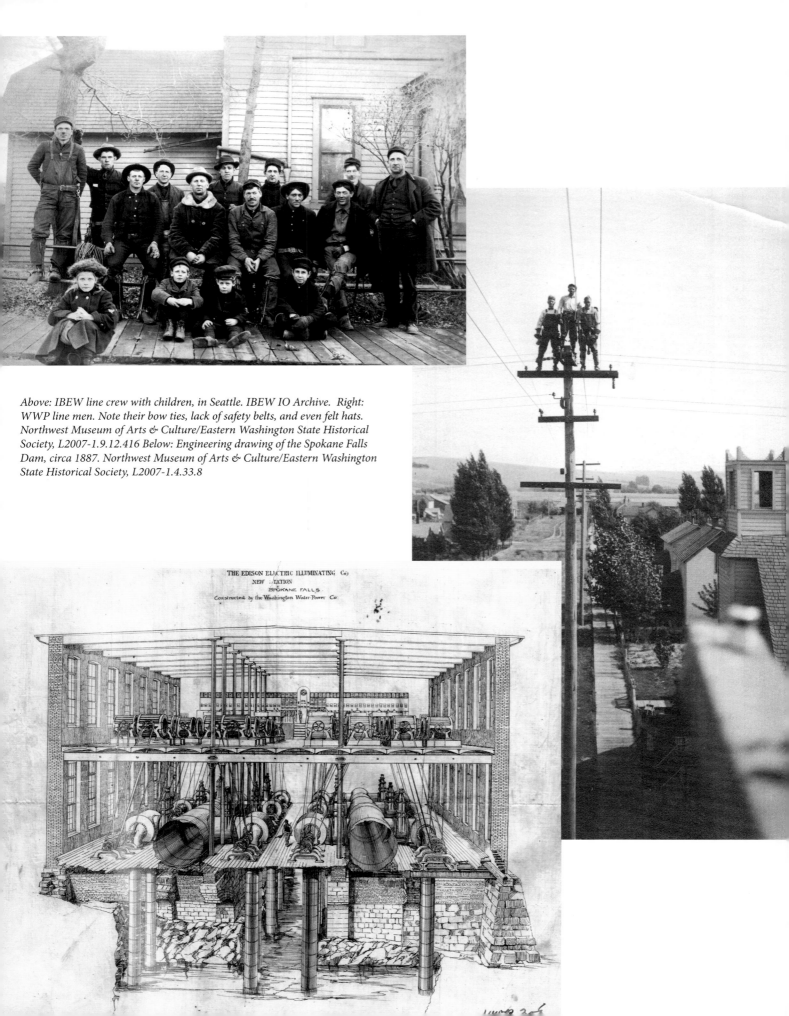

Above: IBEW line crew with children, in Seattle. IBEW IO Archive. Right: WWP line men. Note their bow ties, lack of safety belts, and even felt hats. Northwest Museum of Arts & Culture/Eastern Washington State Historical Society, L2007-1.9.12.416 Below: Engineering drawing of the Spokane Falls Dam, circa 1887. Northwest Museum of Arts & Culture/Eastern Washington State Historical Society, L2007-1.4.33.8

THE EDISON ELECTRIC ILLUMINATING Co.,
NEW STATION
SPOKANE FALLS.
Constructed by the Washington Water Power Co.

Beginning to Unionize
1880–1900

(217.)

—TO THE—

American Federation of Labor

HEADQUARTERS: 21 CLINTON PLACE, NEW YORK CITY, N. Y.

SAMUEL GOMPERS, President.

ANSWERED APR 3 1904

Any number of Wage Workers, not less than seven, who are desirous of forming a Federal Labor Union, or of having their Local, District, National or International Union affiliated with the American Federation of Labor, must fill up this form and forward it, together with Five Dollars as Certificate of Affiliation Fee, to this office, as above, for approval.

(CITY AND DATE) St Louis, Dec 9, 1891

We, the undersigned Wage Workers, believing it to be well calculated to improve our intellectual and social condition, and promote our industrial well-being and advancement, respectfully petition the American Federation of Labor to grant a Certificate of Affiliation to us as representatives of:

Name of Organization _National Brotherhood of Electical Workes of America_

Holding Regular Meetings at No. _____ Street,

in the City of _St Louis_ State of _Mo,_

We hereby pledge ourselves, individually and collectively, to be governed by the Constitution, Rules and Usages of the American Federation of Labor, with the reserved right to preserve the autonomy or self government of our own organization, subject to such rules and regulations as may be made, or are now established in our organization as above named.

Total number of members in Union _____

H. Miller PRESIDENT.

J. T. Kelly SECRETARY.

ADDRESS OF SECRETARY _2210 Washington Av_

NAMES OF APPLICANTS.	ADDRESSES OF APPLICANTS.
1 St Louis, 5221 A. F. of L.	J. T. Kelly Pres.
2 Indianapolis 5536 "	J. Carroll "
3 Toledo 5497 "	M. J. Donohew "
4 Duluth, 5557 "	W. A Warnebe "
5 Evansville 5384 "	E. W. Laberton "
6 Philadelphia 5530 "	E. W. Elliott "
Chicago 5552	T. J. Finnell "

4

1880s

FROM the time of the first telegraph lines, there were men (and men only) who set up the poles and then climbed them to string wire. Telegraph service evolved into telephone service. More poles were set and more wire was strung, and the men who did this work followed it across the United States and Canada.

According to the IBEW International Office's own history, *Carrying the IBEW Dream into the 21st Century,* "enough" telegraph linemen organized to affiliate with the Knights of Labor (KOL) in 1880, becoming part of the KOL's 28,000 members. (The KOL grew exponentially in the next six years, to almost 800,000 members in 1886.)

In 1883 a council of linemen "affiliates," or locals, attempted a general strike against Western Union Telegraph. The strike was unsuccessful and the affiliates were no more.

Linemen made another attempt at unionizing in 1884,
> [T]his time with a secret organization known as the United Order of Linemen. Headquarters for this union was in Denver, and the group attained considerable success in the western part of the United States.
> IBEW, *Carrying the IBEW Dream into the 21st Century,* p. 8

No mention of the United Order of Linemen's presence in the Pacific Northwest has been found in researching this book.

Delivery of electricity imitated the telephone line system, sometimes using the same poles, many times running new courses of power poles and electric lines. The goal was to transmit and distribute electricity to industrial, retail, and residential customers. The first electrical linemen were tough and almost entirely self-taught. There was no system to certify work skills or knowledge. Linemen trusted one another with their lives, and suffered a mortality rate of 50 percent, according to the IBEW's 2005 history, *Carrying the IBEW Dream into the 21st Century.* The linemen who stayed alive were either extremely lucky or had gained enough experience to continue.

Linemen recognized two working conditions they could improve through solidarity: their personal safety and the market for their skilled labor. *Carrying the IBEW Dream* describes seven-day-a-week, twelve-hour days for something like $56 a week. Getting pay commensurate with their craft, or even a standardized pay scale, would come later.

Linemen followed their ever-expanding work from city to city, installing transmission and distribution lines between various sources of electricity and wherever someone would pay to receive electricity.

Opposite: December 4, 1891 NBEW application to the AFL. Signatories from St. Louis, Indianapolis, Toledo, Duluth, Evansville, Philadelphia, Chicago; H. Miller, president, J.T. Kelly secretary. IBEW IO Archive.

Above: WWP linemen on a pole that is suspended from its power lines. Northwest Museum of Arts & Culture/Eastern Washington State Historical Society, L2007-1.9.12.1.

Above: NBEW seal, 1893. IBEW IO Archive.

Right: Knights of Labor seal. Wikipedia, "Knights of Labor."

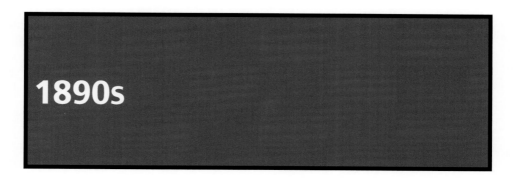

1890s

IN 1890 the twenty-year-old Knights of Labor (KOL) and the four-year-old American Federation of Labor (AFL) were fighting over which organization would represent American workers. The KOL staked its future on worker solidarity across all trades, avowing "That is the most perfect government in which an injury to one is the concern of all." But the KOL's membership had shrunk down to 100,000 members. The AFL started as a federation of separate craft unions in opposition to what they saw as the KOL's challenge to craft-oriented unionism. The AFL affiliates wanted to improve wages and working conditions while avoiding direct political action.

That same year, linemen from across the United States went to St. Louis to run wire and build structures for the St. Louis Exposition, one of the first major public electrical installations in the United States. They compared their various rates of pay and working conditions and decided to organize as a craft union. With the help of Charles Cassel, an AFL organizer, they were chartered as Electrical Wiremen and Linemen's Union #5221, of the AFL, and elected Henry Miller president.

One year later, on November 21, 1891, ten delegates representing 286 members reconvened in St. Louis, above Stolley's Dance Hall, for their first convention. Over seven days they set an ambitious agenda.

> [D]uring their convocation they composed a constitution, established a dues and per capita structure, planned for a death-and-strike benefit, and limited strike activity to only that sanctioned by national officers.
>
> *Encyclopedia of U.S. Labor and Working-class History*, edited by Eric Arnesen, p. 663

Local #5221 adopted a fist grasping lighting bolts as its logo, then elected Henry Miller first grand president and J.T. Kelly grand secretary-treasurer.

The National Brotherhood of Electrical Workers (NBEW) put out the first edition of its national magazine, *The Electrical Worker,* in January 1893. Ever since, each issue has included field reports from locals across the country. Month after month, pages of the early *Electrical Worker* were filled with flowery notices called "resolutions", each saluting the good character of a co-worker killed on the job. Reports also included updates on injured linemen, call-outs to individual linemen wherever they might be, warnings (with physical descriptions) of scabs (non-union linemen), and updates on how much or how little work was available within a local's jurisdiction.

The first NBEW locals in the greater Northwest were chartered, as start-up utility providers in the region jockeyed for service areas and sources of electricity.

Once communities heard about, or in some cases saw, electricity, they wanted it. Some of this desire was to electrify specific mechanical tasks and chores but there was also a sense that electricity, and especially electric light, was modern. As soon as it became even marginally available, demand for electricity surged.

At this point there were no standards for electrical generation, transmission, distribution, or customer service. Each burgeoning electrical utility provided a mix of electrical service depending upon its generating source and location, its service area, its available technology, investment capital, and its market.

Tacoma Light and Water Company, franchised by the Tacoma City Council, provided some lighted streets by 1885, one of the first municipal electrification projects in the Northwest. Precursors of Washington State's current major utilities, Seattle City Light (SCL), Puget Sound Energy (PSE, formerly Puget Sound Power and Light, PSP&L), and Avista (formerly Washington Water Power), were generating electricity by 1892.

Municipal trolleys or streetcars were among the first major electrical uses in the Northwest. Once trolley service existed, many small providers were bought out or combined into larger trolley companies and/or electric companies.

The NBEW's growth paralleled that of the electrical industry. Its official jurisdiction expanded to include Canada, and it changed its name to the International Brotherhood of Electrical Workers (IBEW) in 1899.

IBEW locals represented a variety of work related to expanding municipal services: trolley system workers, the "telephone girls" and the "telephone boys," and linemen. IBEW also represented those who installed and maintained electrical service, which included both linemen and wiremen in early days.

On August 28, 1897, #77 was chartered into the American Brotherhood of Electricians (ABEW) with eight members. (An earlier, very short-lived #77 had been chartered in Detroit, from June 16, 1894 to September 1, 1895.)

Washington State's #77 was given initial jurisdiction of "the greater Seattle area." Receipts from its first meeting were $16, officers were "nominated, elected, and installed," and trustees were assigned to "procure a hall and notify all members as to the date and place of meeting."

During the next two #77 meetings, in March 1898, five more workers applied for membership, the eight existing members paid their per capitas and an additional dollar each for a "Death Assessment," and the thirteen-member local authorized payment of rent for use of a hall.

Local #77's bylaws had been approved by the NBEW's General Secretary by its April 17, 1898 meeting. The membership directed that these bylaws be printed along with working cards and dues cards. More new members were voted in and initiated. The membership also voted to spend $10 for "an entertainment" to be held later that month.

There is little information on exactly what job classifications #77 represented during its early years. Local #77 had some non-linemen members, as well as allowing a sort of "side local," B-77, for the telephone girls, at some point. Part of the difficulty in documenting membership is that few of #77's early minutes still exist.

Records do show that #77 received correspondence from various unions across the country requesting support. The local sent donations when it could. Local #77 correspondents in turn wrote to *The Electrical Worker* to ask other NBEW locals to support non-NBEW utility workers in the Seattle area, such as the "telephone girls." (Much of the information for this period comes from IBEW #73's *Mini History of I.B.E.W. #73*.)

Such active solidarity was due in part to the lack of any legal protection for union activity. Going into the twentieth century there was almost no state or federal law that granted workers collective bargaining rights or even the right to organize. Except for railroad workers, such legislation wouldn't come until the National Labor Relations Act of 1935 (NLRA, also known as the Wagner Act).

Prior to the NLRA, agreements between workers and their employers fell somewhere between a handshake deal and a letter. Workplaces were technically "open shop," which meant each IBEW member had to hold his own when it came to remaining a member of the union. (The term "open shop" was first used in 1904 by laborers trying to block what seemed like an endless wave of immigrants from lowering all rates of pay, according to Murray's *The Lexicon of Labor*.)

Union linemen were vehement, and many times violent, in protecting their work. At the same time, nascent unions could call a strike as they chose, when they chose, targeting a worksite, an employer, or a category of work, knowing that other unionists would boycott.

*Early line crew circa 1900–1930. Local #77 Archive, photo donated by
lifelong IBEW member Jim Glover.*

City of Seattle power pole marker. IBEW IO Archive.

The Saga of Lew Brooks and His Hooks

George Llewellyn Brooks was president of Local #77 during some or all of the Reid-Murphy split, a period in which the national level of the IBEW split into two factions. IBEW's International Offices (IO) records do not indicate any of #77's officers or staff from 1909 to 1912, but do list Brooks as president from 1913 to 1914. Several sources cite Brooks as the driving force behind the 1913 Safety Standards for Electrical Workers (Chapter 296-45, Washington Administrative Code), the first electrical safety law established in the United States and one that served as a model for laws in other states. His nickname is spelled both "Lou" and "Lew," even in documents that seem to come from his son.

The following is from #77's monthly field report to The Electrical Worker, *a month after Brooks' Michigan local sent in word that Brooks had died.*

> We read with sadness of the untimely death of Brother George L. Brooks, a member of Local Union No. 362, Lansing, Mich. Brother Brooks at one time was a member of Local Union No. 77, and a foreman for city light [Seattle City Light]. He and Mrs. Brooks were both active union boosters. Lew was active here in state politics for safety labor legislation.
>
> Brother Brooks, before his death, perfected and placed on the market an adjustable climber for linemen. We have ordered a sample pair. Local Union No. 77 holds that Brother Brooks was among those responsible for the creation of the Safety and Industrial Relations Department in the state of Washington.
>
> We, therefore move that the ELECTRICAL WORKERS' JOURNAL act as clearing house to present a picture of Brother Brooks and Brother Jay Olinger, supervisor of safety and industrial relations. This picture to be placed on the walls of Brother Olinger's office in the state capitol as a gift to the state of Washington to commemorate the services of Brother George L. Brooks. It is further suggested that each outside local order from Mrs. Brooks a sample pair of Brooks climbers....
>
> Frank Farrand
> *The Electrical Worker*, June 1934, #77 report

George Llewellyn Brooks, from his wedding portrait. IBEW IO Archive (from information donated by #77). Opposite: Tool catalogue advertisement for Brooks Hooks, circa 1980. IBEW IO Archive (from information donated by #77).

BROOKS POLE & TREE CLIMBER

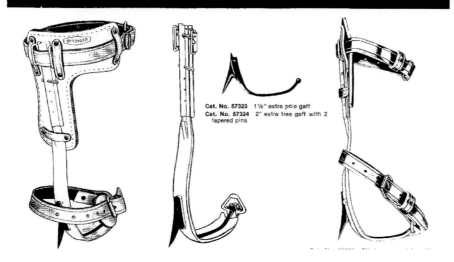

Cat. No. 57323 1 ½" extra pole gaff
Cat. No. 57324 2" extra tree gaff with 2 tapered pins

The most extensive information about Lou Brooks comes from a packet given to the IBEW IO by Dick Rogers while he was Business Manager of Local Union 77, in 1980. It appears that George Llewellyn Brooks's son, George F. Brooks, wrote the following narratives, one of which is signed G.F.B. and dated January 20, 1978.

LINEMAN, LEADER, LAWMAN, INVENTOR

Lineman George Llewellyn Brooks was President of Local #77, International Brotherhood of Electrical Workers at Seattle, Wash. when he, and his appointed committee, drafted a rigid set of electrical safety rules. I.B.E.W. meetings in those days were held secretly up in the mountain forests to avoid conflicts with union-busting "goon squads."

Pres. "Lou" Brooks lobbied at Washington's state capitol in Olympia until his safety regulations were passed by the Legislature (Senate Bill #422) and were enacted into law with Governor Ernest Lister's signature on March 20, 1913. These were the first electrical safety laws established in the United States and they served as a model for those to come in other states.

George "Lou" Brooks, born near Springfield, Ontario, was raised on a farm in northern Michigan. At age 12, Lou quit school (4th grade) and went to work in a lumber camp in Michigan's Upper Peninsula. This was followed by a decade of employment as a brakeman on railroads in Mich. and the State of Washington. In 1908 Lou hired in as a lineman for The City Light Company of Seattle.

Immediately after Lou's Electrical Safety Regulations became law in 1913, Brooks was fired by City Light and soon learned that he was "black-balled" by all utility companies on the West Coast. He was forced to return to his old trade and found employment as a brakeman for The Northern Pacific R.R. out of Auburn, Washington.

In 1917 Lou inherited his father's homestead farm back in Michigan but soon found it necessary to supplement the farm income. He went to work for Consumers Power Co. as an itinerant lineman. In 1923 Brooks was employed by The Board of Water & Light Co. at Lansing, Mich. and soon was promoted to Foreman. His line crew was assigned exclusively to the job of installing Lansing's first traffic signal lights. Brooks was frequently called in for consultation and suggestions by the utility firm's engineers and occasionally he was assigned to duty as the Acting Superintendent.

Here again, Brooks was fired in 1931 as a result of a hassle with management concerning the safety of his crew. Lou then went on WPA and with an inexperienced crew was re-assigned to the same old chore—installing Lansing's traffic lights.

For almost two decades Brooks had been toying with an idea—a design for an improved and safer pole climber. On May 1, 1934, just 50 days after Lou had passed away, his patent for "Brooks Hooks" was granted. The story of the development of the patented design and the problems connected with the early promotion of the new safety climber follows—as told by his son.

The following appears to have been written by George F. Brooks, George Llewellyn Brooks's son, but is unattributed in the packet of documents compiled by Dick Rogers. It is placed here to follow the chronology of events it relates.

> Brooks Hooks
> Patent No. 1956852
> Invented by George Llewellyn Brooks

At age sixteen, one beautiful morning in August 1930, I decided to be an electrical lineman, just like my Pa, George L. Brooks. Donning his "hooks" (lineman's climbers), I walked across Alpha Street and started up a pole.

It was easy to drag the gaff (spur) up the pole six to ten inches at a time, alternately shifting my weight from one foot to the other. In a minute or two, near the top, I paused, surveyed the landscape and checked out the horizon for cowboys and Indians. None were seen anywhere across Shubels' farm, even as far as Sycamore Creek, so I decided to descend.

"Hell's Bells! Woe is me!" The gaffs were tightly imbedded in the pole—I couldn't pull either foot off the pole to step down. I was trapped. Finally, after a five or ten minute battle, I managed to get one foot loose and step down a ways. But then I found it even more difficult to free the other foot from the pole.

After a half hour of struggling, I had gotten down only about six feet, and I was tempted to holler for help. But the embarrassment would have been just too much. So I spent the rest of that forenoon trying to get down, and finally reached terra firma out of breath, exhausted and trembling.

Pa came home for lunch and I told him of my harrowing experience. He laughed, gave me the raspberry, then explained, "You don't pull the gaffs up to get them out of the pole, you bend your knee outward, away from the pole and at the same time, you twist your ankle so that you can pry against it with your foretoe. This way, you break a chip out of the pole, sideways, and the climber is out."

That afternoon I gave it another try, but with the leg irons coming four inches above my knee, instead of down on the calf, my leg was splinted—it couldn't be bent, so that I still couldn't free the gaffs from the pole.

After supper, I asked Pa if they made different length climbers for tall or short men. He replied, "I once knew a lineman, just five-foot-two, while another was nearly seven foot. Still, they both had to use the same climbers; but I'm planning to make mine adjustable in height some way."

For two decades or more, Pa had been tinkering with a model for climbers with removable and replaceable gaffs. Out in the Seattle-Tacoma area, back in 1912, Pa twisted his ankle so badly that he had to use crutches for a while. The crutches were so short that Pa had to bend over to use them. Ten-year-old Marguerite, my Sis, asked why they didn't make them adjustable like telescopes, so that they would fit anybody. It never occurred to Pa to run to a patent attorney, so later on somebody else made a bundle. But Pa realized instantly that the same principle was needed to improve the lineman's climbers.

Entry from #77's 1914 dues book, contemporaneous with Brooks. J.J. Wilson, age 27, paid $1.50 in dues. All member payments for August and September 1914 were marked "came from 77 formerly McNulty." Local #77 Archive.

The following section is signed by George F. Brooks, son of George Llewellyn Brooks', on January 1, 1978. It is placed here to follow the chronology of events it relates.

In September 1930, I enrolled in an eleventh grade drafting class at Lansing Eastern High School. One of my first plates (drawings) was to illustrate cross-sections of various structural forms used for metals, such as: flat (strap); L (ell or angle); U (channel); T (tee); Round (or tubular); Hexagon; Square (box) and so on. Under each illustration we had to letter the coefficient figure, which we found in the Engineer's Handbook.

This numeral showed the relative strength and rigidity of each of the structural forms as compared to the same amount of metal made up in strap form. I noticed that the channel and T forms were more than double in comparable strength to the flat strap formation.

I also remembered how heavy those flat steel lineman's climbers were after lugging them around for only a short time. It gave me an idea, so I took my drawings home. That evening I asked Pa, "Don't those climbers get pretty heavy when you wear them all day?" Pa replied, "About 4:30 p.m., they weigh sixteen ton!"

Then I showed him my drawing, pointing out that climbers would be just as strong, using half as much steel, if they could be made in channel or tee formation.

His eyes lit up as he blurted, "Yup, and that ain't all!! Think I've got the answer to a problem that's had me stymied for years!"

Next day, Pa came home with a piece of half-inch T-iron and a piece of sheet metal which he had bent to fit tightly around the T-iron, so that the sheet metal would slide along the iron. He had drilled four holes, spaced about an inch apart, in the vertical rib of the T-iron. Likewise, he had drilled a series of holes in the sheet metal. Aligning various sets of holes, and pinning them with a cotter key, he showed me how to lock the sheet metal slider at various positions, 1/4 inch apart, for a total distance of three inches.

Pa said, "Now, any lineman, regardless of his height, can wear his hooks just where he likes them best—low on his calf, or high up to the knee." He handed me his new telescoping gadget, along with his ten-year-old aluminum model with the replaceable gaff. He said, "I want you to ask your Drafting teacher if you can make a drawing of this."

My instructor, Mr. T. K. Clark, agreed, saying that it would give me a good exercise in the use of the French curve, and that he would give me credit for five drawings.

Engrossed as I was in this project, I proceeded to get ten drawings behind the rest of the class. But when the drawing was completed, Pa took it to a patent attorney, Samuel H. Davis, who executed an official "Evidence of Conception" form, which would serve as temporary protection while a patent was pending. Ironically, the patent was issued, effective May 1, 1934—just two months after Pa died.

During the three-year interim that the patent was pending, Pa spent most of his spare time traveling throughout the midwest, getting manufacturing ideas, bids on the dies, and searching for an "angel" willing to invest up to $500, or even $1000, for the dies. In 1931–32, this kind of money would nearly buy the Waldorf-Astoria—but Pa pressured one man until he finally gave in out of sheer desperation.

Mr. J.W. Wolford, president and owner of Melling Drop Forge in Lansing, Michigan, and also the first president of the newly established Bank of Lansing, agreed to pay for the dies out of his own pocket, mostly because of his die-sinkers and hammer-men at Melling Drop Forge, who had no other work.

Pa recruited a crony, a Mr. Smith, retired lineman, to go on the road, taking orders for the climbers. So "Smitty" bummed the nation for the next three or four years, living on handouts and a bed provided by some generous lineman, keeping as his sales commission the $2 sales deposit on each order he took. Most of these original buyers never dreamed that they would be waiting for years to get delivery.

During these early, and difficult years of production, my brother, "Bud" (Clifford Llewellyn Brooks) did an excellent job of handling the shipping, bookkeeping, billing and, especially, in appeasing the buyers' complaints about non-delivery of orders. He did get some help from us kids with the shipping details.

Then, the patent rights were sold to Melling Drop Forge on a royalty basis. Later, Melling turned the distribution over to Mine Safety Appliances Corporation, Pittsburgh, Pennsylvania, who sold them throughout the United States, Canada, Mexico, and Central and South America. For the past few years, the climbers have been distributed exclusively by Stringer-Brooks Corporation, Browning Street and Highway 50, Lee's Summit, MO, 64063.

"Smitty", the initial sales staff and promoter of "Brooks Hooks", died with his boots on, somewhere out in the Southwest, no doubt shortly after having some suds with the $2 commission from his last sale. He's buried out on a lone prairie somewhere, God rest his soul.

While Smitty sowed the seeds, my brother Bud cultivated the sparse crops of sales, until Mine Safety Appliances spread the "Brooks' Hooks" from the Yukon Territory to Argentina and Chile. One way or another, almost all of the Brooks family got their hooks into "Brooks' Hooks", but they would never have gotten off the ground (and up a pole) if it were not for the faith and foresight of Mr. J.W. Wolford, and Pa's old sidekick Smitty.

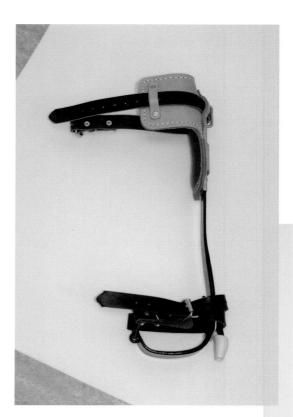

Gaffs, or climbing hooks, have evolved over the years.
Left: Gaffs circa 2011. Note leather padding and plastic gaff cover. Local #77 Archive.
Below: Unusual gaffs from before the era of Brooks Hooks. Metal arm was swung around pole. IBEW IO Archive.
Right: Gaffs used by George E. "Ed" Noyes while a lineman with PSP&L. Noyes worked for PSP&L from the mid-1940s until 1970. Loren Noyes Collection.

Opposite: George L. Brooks's certificate of patent for "linemen's climbers" May 1, 1934. IBEW IO Archive (from information donated by #77).

1956852

THE UNITED STATES OF AMERICA

TO ALL TO WHOM THESE PRESENTS SHALL COME;

Whereas   GEORGE L. BROOKS,

of

Lansing,   Michigan,

PRESENTED TO THE Commissioner of Patents A PETITION PRAYING FOR THE GRANT OF LETTERS PATENT FOR AN ALLEGED NEW AND USEFUL IMPROVEMENT IN

LINEMEN'S CLIMBERS,

A DESCRIPTION OF WHICH INVENTION IS CONTAINED IN THE SPECIFICATION OF WHICH A COPY IS HEREUNTO ANNEXED AND MADE A PART HEREOF, AND COMPLIED WITH THE VARIOUS REQUIREMENTS OF LAW IN SUCH CASES MADE AND PROVIDED, AND

Whereas UPON DUE EXAMINATION MADE THE SAID CLAIMANT is ADJUDGED TO BE JUSTLY ENTITLED TO A PATENT UNDER THE LAW.

NOW THEREFORE THESE Letters Patent ARE TO GRANT UNTO THE SAID

George L. Brooks, his heirs   OR ASSIGNS

FO THE TERM OF SEVENTEEN YEARS FROM THE DATE OF THIS GRANT

THE EXCLUSIVE RIGHT TO MAKE, USE AND VEND THE SAID INVENTION THROUGHOUT THE UNITED STATES AND THE TERRITORIES THEREOF.

In testimony whereof, I have hereunto set my hand and caused the seal of the Patent Office to be affixed at the City of Washington this first day of May, in the year of our Lord one thousand nine hundred and thirty-four, and of the Independence of the United States of America the one hundred and fifty-eighth.

Attest:

G. P. Tucker

Law Examiner.

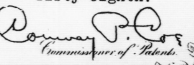

Conway P. Coe

Commissioner of Patents.

Clockwise, from above left: SCL's Skagit, or Diablo Dam, Incline railway. University of Washington Libraries, Special Collections, UW27566. SCL workers guiding rotor into place on power-generating unit 2, Gorge Dam Powerhouse, February 25, 1924. University of Washington Libraries, Special Collections, UW27568. SCL crew riding incline railway car to work. University of Washington Libraries, Special Collections, UW27565. Gravel for Skagit River hydro project transported by incline railway. University of Washington Libraries, Special Collections, UW27566. Tourists riding the incline. Lee Look Collection.

Following the Electricity
1900–1949

Above: Puget Power linemen in front of Seattle lineroom,1910. (This is now Seattle's Medical-Dental Building.) Top row L-R: William Dick, E.M. Bird, Snow, Roy Heckathorn, John Ridley, George Stocks, Jones, Carl Duley, McClaffety. Bottom row L-R: Fred DeSylvia, George Jackman, C.E. Bird holding Rags the dog, Dan Trink, Jim Perry, Hamm. University of Washington Libraries, Special Collections, UW35551.

Lower left: Portraits of Charles A. Stone and Edwin Webster. (Stone and Webster were and their company remains a major utility contractor.) University of Washington Libraries, Special Collections, UW35568.

Lower right: map showing locations of PSP&L's hydroelectric and "steam-electric" generation plants. University of Washington Libraries, Special Collections, UW35566.

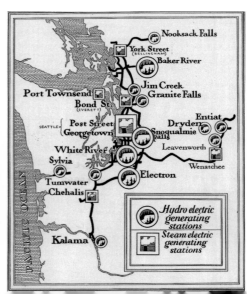

A New Century

CITIES generally issued franchises to private corporations to provide electricity and other utilities such as water and transportation. These corporations then provided limited utilities to limited areas of a city, depending upon technological limits, geographic distances and topography, and which residents were willing to pay. (Many cities still franchise investor-owned utilities for services within their corporate limits.)

In few cases did anyone outside of an urbanized area have access to electricity unless they had onsite irrigation.

As Paul Hirt writes:

> Corporations holding franchises to provide lighting, power, and transportation services to municipalities made up a special category of business called a "public service corporation" or, increasingly, a public utility. (The nomenclature can be confusing because these are usually privately owned, for-profit companies. A "public utility" does not necessarily denote a publicly owned utility.)
> Paul W. Hirt, *The Wired Northwest*, p. 62

The first franchise agreements relied upon this mix of "public" and "private" terminology and practices. Public and private financing, legislation that sometimes groups and sometimes separates publicly and privately owned utilities, and varying degrees of cooperative and competitive utility operation have remained part of the Northwest's utility system ever since.

Once alternating current (AC) provided the technology to transmit electricity a significant distance from its generating source, the original assortment of small utility corporations consolidated and bought each other out in a flurry of deal-making that makes a game of Monopoly seem like child's play.

This rush to consolidate was driven in part by the need for capital—cash to buy machinery for generating and transmitting electricity, cash to pay workers to install equipment and run transmission lines, and most importantly, cash to purchase and develop hydropower sites. Hydropower was *the* source of electricity in the Northwest.

Few of the inventors and engineers who knew how to make electricity work had access to the scale of financing hydropower requires, so they negotiated with big investors from the East Coast. Some of the biggest of these investors were the patent holders and/or sole manufacturers of the very machinery of electricity: Edison Electric (which through mergers would become Consolidated Electric, or Con Ed), General Electric, and Westinghouse.

In some cases the consolidation of utility corporations improved the cost, availability, and consistency of municipal electrical and trolley service. In many situations it did not.

Utility consolidation in the Northwest was part of a bigger wave of merging financial investment in utilities. Nationally and internationally, corporations were interested in controlling the burgeoning business of electricity. Their investors were even more interested in quick profits from arbitrage in the volatile utility market. (Arbitrage is the sale of corporate securities between markets for immediate profit.)

As corporations combined and small companies were absorbed by larger companies, IBEW locals struggled to maintain representation of their existing members. Until after the turn of the century, the IBEW did not classify "inside" work separately from "outside" work. For example, in July 1900, #73's minutes mention representing Washington Water Power (WWP) workers, including "armature winders."

In January 1902, shortly after the IBEW did create separate classifications for wiremen (inside electricians) and linemen (outside workers), #77's press secretary, Jack Cameron, wrote:

> I have to state that the linemen and insidemen have disbanded and formed two separate Locals, the linemen retaining the old charter of No. 77, and the insidemen were granted a new charter known as Local No. 217. We have elected Brother Leedy business agent, and there is no doubt he is doing lots of good for the two locals.
> *The Electrical Worker,* January 1902, #77 report

Take note: both locals began by sharing one business agent. Further history of Local #217 is murky, with conflicting accounts appearing in different records.

Meanwhile, the IBEW's International Office (IO) began to affiliate what could be called second-tier or sister locals that were assigned "B" as part of their duplicate number. During the early 1900s "B" locals were usually women's locals, representing assembly workers on the East Coast and telephone company operators and office workers throughout the United States.

Local #77 wrote to the IO saluting the way B-77 members ("the telephone girls" in this case) held their own during a walkout. Local #77 and other linemen locals also wrote in to describe the significant leverage their sister locals had with utility companies, and how their sisters used their leverage to support the linemen. From very early on #77 members could see that organizing *all* the employees of a utility company and coordinating union activities across job classifications would be critical to worksite representation for its linemen.

Although source material is sketchy, it seems that by about 1906 #73's membership took a more contentious approach to the separation of wiremen and linemen job classifications.

Dissension within #73 between its inside and outside members began to appear in #73's minutes just before a strike in March 1906, when linemen made a motion to withdraw from #73. From May 1906 until March 1907, #73 members were on strike at Washington Water Power's Natatorium. (The Natatorium was an amusement park and swimming pool miles outside of Spokane that WWP ran to secure ridership for its trolley system.)

The linemen's motion to withdraw from #73 failed. The linemen presented a withdrawal petition just after the Natatorium strike was settled, in July 1907. At the same meeting wiremen's dues were raised. Two months later, minutes record a report that "all the WWP men" were in #73.

Within the IBEW a bigger split was growing, based on the differences between inside wiremen and outside linemen. What *Carrying the IBEW Dream* calls "a large percentage of the Brotherhood" seceded in 1908 and attended a separate international convention. This dissenting group elected its own officers, President J.J. Reid and Secretary J.W. Murphy. It seems that most of the linemen's locals and all of the Pacific Northwest locals were part of the Reid-Murphy group.

The passage of a hundred-plus years has only begun to soften the bitterness of this division, and the history of the Reid-Murphy faction is not readily available. Even now the IO's *Carrying the IBEW Dream* makes clear which side of the split it was on, declaring that "Frank J. McNulty and Peter W. Collins remained the true officers of our Brotherhood."

After both factions took legal action to get control of the IBEW's funds, Samuel Gompers, president of the AFL, recognized the McNulty-Collins faction as the "legitimate Brotherhood."

The Industrial Workers of the World (IWW, or Wobblies) had taken up the industrial model of organizing from the floundering Knights of Labor (KOL) and was growing. The IWW challenged the AFL model of craft unionism, arguing that "an injury to one is an injury to all." Its view was that the working class needed to organize as a whole, not by trade, and it called the AFL the "American Separation of Labor" as did another anti-AFL union, the Western Federation of Miners. At the end of the nineteenth century, historian Howard

Zinn estimated there were two million union members (one in fourteen workers), with 80 percent of that union membership in the AFL (*A People's History of the United States*, p. 328). However, the IWW allowed and sometimes encouraged its members to hold other union cards, so overall membership numbers are unclear.

The IWW's significant regional presence is most renowned in forestry and timber-related industries and on the docks, but there were Wobblies throughout the Pacific Northwest labor community as it staged large general protests.

In 1909 IWW organizer Elizabeth Gurley Flynn challenged a Spokane ordinance that limited public speaking and was arrested. Huge crowds showed up and asked the city to arrest them, too. More than 500 were arrested, and four people were killed during the protest. Aberdeen, Washington saw a similar protest the same year.

Local #73 does not mention the Spokane protest in its minutes. It was focused on other matters. Local #609 had affiliated with the IBEW only a few months before the Reid-Murphy split and relations between the two locals were not cordial.

Local #77 had more cordial relations with other Seattle locals, reporting in May 1908, as the Reid-Murphy split was beginning, that it had held its first joint meeting with #217 and #202.

> These joint meetings are to be held monthly with the officers of the different Locals presiding in turn and are for the purpose of drawing the members of the Brotherhood in this city [Seattle] closer together, creating a more friendly feeling among the members and to cuss and discuss such topics as are of interest and benefit to the membership in this city, together with such viands and liquids as the committee sees fit to furnish.
> *The Electrical Worker,* May 1908, #77 report

There is no record in the few notes on #77's activities during the Reid-Murphy split as to whether additional joint Seattle meetings were held.

By May 1910 #73 had "divorced" its linemen, and inside work was hard to find. The local reported to *The Electrical Worker:*

> Work is fair here, but there isn't but half of it union, at this writing, though we are working hard to get the rest of it. But there isn't a possible chance for any more men this summer, as a great many of those who have come in have either loafed most of the time, or found that the town was overcrowded with narrowbacks [wiremen], and drifted on.
>   Don't interpret this as "high board fence" we never had one, and never will. It is just facts.
> *The Electrical Worker,* May 10, 1910, #73 report

While regional IBEW locals continued to fight over their membership and jurisdictions, Washington State passed two laws that greatly affected workers.

The first, Industrial Insurance, Revised Code of Washington (RCW) 51, one of the nation's first workers' compensation laws, passed in 1911. Its first section (51.04.010) states:

> The state of Washington… declares… sure and certain relief for workers, injured in their work, and their families and dependents is hereby provided regardless of questions of fault and to the exclusion of every other remedy, proceeding or compensation, except as otherwise provided in this title; and to that end all civil actions and civil causes of action for such personal injuries and all jurisdiction of the courts of the state over such causes are hereby abolished, except as in this title provided.

Workers gave up their "right to sue" regarding work injuries (and deaths) in exchange for a system of state-determined compensations, to be administered within a new department, Labor and Industries. This system of compensation was not initially funded adequately to provide anything but small token payments.

Local #77 was the major player in the passage of another state law, Safety Standards for Electrical Workers (Chapter 296-45, Washington Administrative Code).

According to a biography of George L. Brooks, #77's president during some or all of the Reid-Murphy split, Brooks was the driving force behind this law. Brooks and other members of #77 reportedly drafted "a rigid set of electrical safety rules" during #77 meetings, and then took them to Olympia in 1913.

> Pres. Lou Brooks lobbied at Washington's state capitol… in Olympia until his safety regulations were passed by the Legislature (Senate Bill #422) and were enacted into law with Governor Ernest Lister's signature on March 20, 1913. These were the first electrical safety laws established in the United States and they served as a model for those to come in other states.

Upper left: SCL blacksmith shop. Local #77 Archive.
Upper right: Charlie Thimm, SCL blacksmith, 1958.
Gary Moore Collection. Left: November 1908 wedding
day photograph of Suzanne Brunner's grandparents,
George S. Brunner and Josephine Geist Brunner. Her
grandfather is wearing his WWP or Spokane Street
Railway Company street car driver's uniform. (Brunner is
#77's assistant business manager in Spokane.) Suzanne
Brunner Collection. Below: A Spokane Street Railway
Company streetcar. Avista Archive.

Dick Rogers (#77 Business Manager), George "Lou" Brooks information presented to the IBEW International Offices, 1980.

This law (Safety Standards for Electrical Workers) was adopted in 1913, the same year the Reid-Murphy split within the IBEW came to an end. IBEW reunification was put to a national vote in late 1913, and passed. The April 1914 issue of *The Electrical Worker* ran a full front-page article on the tentative agreement to reunify:

The members of the Brotherhood anxiously awaited the results of the agreement entered into between the Brotherhood and the Pacific District Council of the unaffiliated organization of Electrical Workers last December [1913], and we are pleased to report that notwithstanding the fact that the General Officers of the seceding organization [the Reid-Murphy faction] did everything in their power to prevent favorable action on the part of their members, approximately five thousand of their former members have reaffiliated with the Brotherhood.

The following is a list of Locals reaffiliating at time of going to press...

Nine Washington State locals were listed: #77, Spokane's #73 and #609, two locals #217 (one in Seattle and one in Port Angeles), #458 (Aberdeen), #523 (North Yakima), #556 (Walla Walla), and #574 (Bremerton).

It took an on-the-ground organizing sweep by the IBEW to bring more wayward locals and members back into the IBEW fold. Organizer W.W. Morgan reported in November 1914:

In company with committee from 77 we met committee from No. 3 of the unaffiliated Brotherhood to arrange satisfactorily the plan of affiliation with 77 as per the tentative agreement....

On the 18th and 19th I was in Seattle and talked with quite a number of No.3 members and everyone which I had spoken to appears to be satisfied at the settling of this controversy as the proper thing.
The Electrical Worker, November 1914, organizing report

Local #3 seems to be the Seattle local's number within the Reid-Murphy faction.

Work was slow for #77 members the following winter.

Editor:

A few lines to let the Brotherhood know that 77 is in the running, but must report the same as nearly all other locals a long list of unemployed members. There is absolutely no construction going on around here if any, in the northwest.

The Pacific Telephone Co. have the least number of men at work at this time, for a number of years previous. The City Light have only a little better than half their men working and with the Stone Webster interests on the unfair list, there is no show to get work at the trade, in fact, it is very hard to get any kind of employment.

In spite of these hard times this local has increased its membership over 100% since last July, due mostly to the untiring efforts of the business agent and the great assistance of the grand office. The membership as a whole, sees no difference, as was anticipated before we accepted the tentative agreement, from the present routine and workings of our local and grand office affairs.

The entire Pacific Coast district has shown a great improvement so far as organization is concerned, from the report of District Council President Morganthaler, who is in town for a few days, working on the violations of the Pacific Telephone & Telegraph Co. He reports that there is a large list of unemployed in all the coast towns, so the Brothers who contemplate coming out in this part of the country, would do well to get in communication with the different locals before starting out.

Hoping that the next time this local has an article times will have 'picked up' and we will all be working.
Yours fraternally,
Chas. Cross,
Press Sec. L. U. No. 77.
The Electrical Worker, February 1915

Grace Palladino, in *Dreams of Dignity, Workers of Vision,* states that going into World War I the IBEW was ready and willing to take on other unions for what it considered its territory. The IBEW claimed "electrical work for electrical workers" in the August 1916 *Electrical Worker,* and identified union carpenters, theater employees, railroad signalmen, and operating engineers as usurping its jurisdiction. Because of a significantly increasing demand for skilled inside work, the IBEW was in a position to take jurisdiction of related work. However, there does not seem to have been a surplus of outside work, at least not in Washington State. Here, lots of workers were having hard times, and some went on strike.

In late November of 1916, some 300 Wobblies gathered in Seattle and headed by boat to Everett to support striking shingle workers. They were met by the Snohomish County Sheriff and more than 200 vigilantes. As many as twenty people were killed (five to twelve of them Wobblies) and close to fifty were wounded. The Governor of Washington sent militia to Seattle and Everett to "maintain order." The Wobblies made themselves scarce, but seventy-five of them were arrested when they returned to Seattle.

Above: PSP&L power pole on the Seattle waterfront, circa 1920. University of Washington Libraries, Special Collections, UW35556. Opposite page, top: Baseball field at Natatorium Park. Avista Archive. Opposite page, center left: Unknown amusement fair with storefront featuring "this hotel cooking by wire." Avista Archive. Opposite page, center right: Natatorium Park's swimming pool. Avista Archive. Opposite page, bottom: Seventh annual convention, Pacific District Council No. 1, Third District IBEW, Seattle, 1910. IBEW IO Archive.

Bringing Power to the People

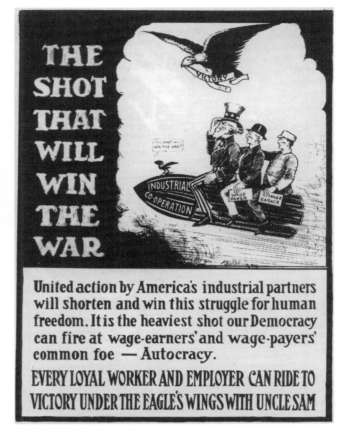

Above left: Open manhole with toolbox, with PSP&L logo on toolbox and protective fencing. University of Washington Libraries, Special Collections, UW35567.

Above right: WWI-era cartoon promoting "united action by America's industrial partners," with Uncle Sam, the wage payer, and the wage earner riding a bomb. IBEW IO Archive.

Below: Worker and generator at Washington's Lake Chelan Hydroelectric Project, circa 1930. Avista Archive.

World War I

GOVERNMENT and vigilante violence against the International Workers of the World (IWW) continued through World War I, even as the Wobblies secured a major labor victory in 1917. The Lumber Workers Industrial Union (LWIU) #120, part of the IWW, struck and won an agreement that led to an eight-hour day for its members. (Interestingly, an IBEW telephone local in Boston had negotiated a contractual eight-hour day in 1913. It would not be until 1937, with the adoption of the Fair Labor Standards Act, that federal law would mandate an eight-hour day for workers in general.)

The U.S. entered World War I in April 1917, and its utility providers struggled to meet an increasing demand for electricity. Factories had been relying upon onsite, coal-driven power. Increased pressure on coal supplies, particularly from munitions factories, led both manufacturers and utility providers to switch to electricity, some generated onsite and some transmitted from hydroelectric or coal-burning electric plants.

When the demand for electricity increased in manufacturing areas like the East Coast, linemen in that region used the growing demand for their skills to improve their working conditions and get union recognition:

> [Linemen] made good use of their temporary economic advantage to sign trade agreements, often for the first time, and to secure the closed or preferential union shop, the Saturday half-holiday, time and a half for overtime, and the eight- or nine-hour day.
> Palladino, p. 103

The IBEW also secured wartime gains for its manufacturing, railroad, and shipyard workers, through direct action.

IBEW activists were jailed during violent strikes in Chicago, at Commonwealth Edison, and in Atlanta, at Georgia Railway and Power Company. The pages of *The Electrical Worker* during World War I are filled with indignant rhetoric asking why the U.S. was fighting for democracy abroad and locking up its citizens at home.

While private employers used the courts to go after IBEW activists, the federal government took over the railroads in late December of 1917, and initiated a policy that recognized collective bargaining by unions. IBEW almost immediately enrolled 90 percent of the electrical workers on the railways, according to Palladino. Similar gains were made at the shipyards.

In April 1918, President Wilson created the National War Labor Board (NWLB) to arbitrate labor disputes in war-related industries. The NWLB did not support strikes, but it did recognize workers' right to organize and collectively bargain, the eight-hour work day, and equal pay for women. This gave the IBEW, in concert with the International Association of Machinists (IAM), what they needed to take on General Electric at its Schenectady, New York and Lynn, Massachusetts plants. The NWLB ruled in favor of the IBEW and IAM in October 1918.

Above: Local #77 members celebrate Labor Day, September 5, 1919. Local #77 Archive.

Below: PSP&L officials line up near the Baker River to commemorate twenty-five years of operation operations, circa 1925 "entertaining Mr. Stone and Mr. Webster." Stone and Webster's corporation (of the same name) has been a major engineering and construction contractor and investor in utilities since the 1880s. University of Washington Libraries, Special Collections, UW35561.

Giant Power and Super-power

ALMOST as soon as World War I ended, labor went on strike as never before. One of the largest and most significant strikes of this period was the Seattle General Strike in February 1919.

Because their work was considered central to Seattle's general health and safety, the linemen of #77 were among the few workers authorized by the strike committee to keep working during the strike.

Local #77's streetcar operators did join the general strike, although they were the first to return to work. "Six streetcars began operating the morning of February 8," according to Historylink's web article on the strike.

In April 1919, two months after this strike, #77 had its charter revoked by the International Office (IO). There are only rumors as to why. Some local "greybeards" (longtime workers approaching retirement) heard stories of #77's business manager at the time fighting with someone from the International Office, possibly a district council representative. Whatever the reason, a new Seattle local was chartered a few days later. Local #944 was a "mixed trade" local. (Local #944 retained the International's recognition, replacing #77, from 1919 into 1925.)

While Seattle locals were straightening out their recognition with the IO, the International was focused on its East Coast locals. Things went poorly for West Coast telephone workers when IBEW called a general telephone strike in June 1919. A settlement was reached on the East Coast, but workers at Pacific Telephone and Telegraph (in California, Oregon, and Washington) failed to ratify the deal the IBEW had negotiated. They were left with no contract and immediate demands to join company unions.

Meanwhile, the Washington State Legislature seems to have taken a mixed approach to union membership after the Seattle General Strike, passing RCW 49.36.010 the same year. This bill establishes the legality of trade unions in Washington State:

> It shall be lawful for working men and women to organize themselves into, or carry on labor unions for the purpose of lessening the hours of labor or increasing the wages or bettering the conditions of the members of such organizations; or carry out their legitimate purposes by any lawful means.

However, the legislature had also passed a law in 1918 which made it illegal to "advocate crime, sabotage, and violence as a means of accomplishing political or industrial reform." Upheld by the Washington Supreme Court in 1920, this law was particularly applied to the IWW, as were others like it (these were known generally as criminal syndicalist laws, and were passed by half of the states shortly after World War I).

At the federal level, unions faced additional legal prosecution. The Justice Department's Palmer Raids targeted "Bolsheviks" and Communists in 1919 and 1920 and stopped only when these raids became so excessive that the Labor Department quit cooperating with the Justice Department.

The AFL unions faced an additional political challenge: the Open Shop Movement. (A workplace is open shop when there is a union but union membership is not required.) Industrialists feared that the National War Labor Board (NWLB), in granting better terms and stronger unions for workers, might encourage a social rebellion on a par with the Russian Revolution. Corporate leaders also used such fears to tar the labor movement as un-American.

Beginning in about 1920, the National Association of Manufacturers and other business interests sponsored what they called the "American Plan," which equated open shops with patriotism and union shops with Bolshevism. Supporting manufacturers pledged they would not negotiate with union representatives.

The IBEW took on the American Plan and the Open Shop Movement within its industry by meeting with the Conference Club, a group of electrical contractors. By March 1919, the IBEW had signed a Joint Declaration of Purpose with both the Conference Club and the National Association of Electrical Contractors (NAEC, later to become NECA). NAEC and IBEW ratified an agreement to form the Council on Industrial Relations (CIR) in 1920. The CIR provided a way for the IBEW to systematically meet with employers and management to settle labor disputes, and recognized the IBEW as the union representing electrical workers.

In Washington State the IBEW faced different issues. Although the CIR in 1920 technically took over jurisdiction of disputes with electrical contractors, in reality it was instituted to deal with building construction (inside work), not utility construction (outside, or linemen's work) which is what dominated in the Pacific Northwest. Also, most linemen worked directly for the utilities, not for contractors. In addition, there was virtually no manufacturing in Washington State that fell under IBEW's jurisdiction, so there were no IBEW-related factory strikes here, nor much, if any, opposition from the National Association of Manufacturers.

There was another critical regional difference within the IBEW. Hydropower dominated the West as its primary source of electricity, while coal was the primary generating source in the East. This meant there was a race to get control over potential hydropower sites in the West. The same corporations that had control and/or ownership of electricity in the East, both its coal fields and its generation, were gaining control of the future utility industry in the West.

Just as when the telegraph industry developed, a few investment groups and corporations gained control of electric power production, distribution, operation, and management of utility systems across United States. By World War I the private electrical utility industry in the United States had settled into a mostly comfortable milieu of state licensed monopoly enterprises, with episodes of bitter competition between municipal utilities and public (nonprofit) power advocates.

> [Utilities operated as] a business government partnership to provide profits for investors while advancing the interests of consumers, cities, states, and the nation.
> Hirt, p. 131

Eastern utility and manufacturing corporations and their holding companies faced off with the IBEW. (A holding company owns the stock of other companies, and usually does not produce its own goods or services.) Many of these same entities already owned the means of electrical generation in Washington State.

> [E]lectricity accelerated trends toward the increasing concentration of wealth and power in the hands of an elite class of national and international business leaders and financiers, what Alan Trachtenberg has called "the incorporation of America."
> Hirt, p. 95

In response to financial consolidation across many industries, Congress began to create regulatory agencies to protect the public's interests. One of these, the Federal Power Commission (FPC), was created in 1920 by the Federal Water Power Act. The FPC was mandated to create and enforce unified governance of hydropower, replacing and overriding various state regulations.

Until about 1930, the FPC did little more than issue long-term licenses for private hydropower. The FPC had neither the staffing nor the wherewithal to take on the phenomenal interstate consolidation of electric power holding companies.

> By 1924, having superimposed themselves on the large and small utilities, the large companies that dominated the utility industry had become holding companies... In that year, holding companies controlled two-thirds of the generating capacity of the industry; seven holding company groups controlled forty percent, and other holding companies controlled twenty-five percent.
> Thomas P. Hughes, *Networks of Power,* pp. 392–3

Individual states began to lobby for strict, federally coordinated and controlled leases on public power systems, be they hydropower or coal-generated. Governor Gifford Pinchot of Pennsylvania led the movement, which was called the "Giant Power Movement." The "Super-power Movement" was its antithesis, working to maintain unregulated electrical generation. Pinchot explained the difference in 1925.

> "[G]iant power and super-power are as different as a tame elephant and a wild one." The first was a "friend and fellow worker of man, the other, at large and uncontrolled, may be a dangerous enemy." The main object of giant power, Pinchot said, was to "assure better service and vastly cheaper rates to the consumer... through effective public regulation." The chief idea was "not profit, but public welfare." Super-power, on the other hand, was "the interchange of small quantities of surplus power at the ends of the distribution wires of each system." The main object of the super-power idea was "greater profit to the companies."
> "Pinchot's Plan to Tame the Elephants," *Literary Digest* 84 (March 3, 1925): 13, as quoted in Philip L. Cantelon, "The Regulatory Dilemma of the Federal Power Commission, 1920–1977"

The consolidated corporate push behind the Super-power Movement was clear to the IBEW and its members. But the IBEW had experienced minimal success in taking on these corporations and their predecessors in the telephone industry. In addition, the union faced challenges from the left within the labor movement. IBEW's national leadership chose to focus on making sure the exploding electrical industry could stay in IBEW's jurisdiction rather than taking on corporate ownership of the utility industry.

At the 1925 IBEW Conference in Seattle, IBEW President J.P. Noonan forcefully outlined the IBEW's strategy for backing the Giant Power Movement. Before he took on "the power industry," he reminded the membership of the continuing threat from the Open Shop Movement. He first explained to the membership why he would not go into specific details of the IBEW's strategies:

> [P]lans should not be... detailed in a report of this nature, as we do not believe it to be good policy or tactics to set them forth in documents open to the public, as the incoming officers, in order to obtain any measure of success with any plan agreed upon, must enter upon such campaign without the plans of it being advertised to the world at large and met by the opposition with counter-movements before any effective action will have been taken.
> "President Noonan's Remarks," *IBEW Convention Proceedings 1925*

The still tender reunification of the IBEW meant Noonan's "opposition" was not only from outside the IBEW. Noonan then explained that although he saw Giant Power as simply a more regimented version of Super-power, the IBEW needed to focus upon maintaining a unionized labor force, no matter who owned the utility industry.

> The so-called giant power project has been widely heralded throughout the United States, and a great deal of time, energy and money has been devoted to propaganda purposes to make the people ready for the acceptance of the programme. It is the intent and purpose to build large, up-to-date power producing units on waterpower sites and in the coal mining districts....
>
> Hundreds of millions of dollars' worth of bonds have been issued by power companies bringing about consolidation, and we have to believe that, what the American Telephone and Telegraph is to the telephone industry today, the Electric Bond and Share Company will be to the power industry in a comparatively few years....
>
> An investigation showed the Electric Bond and Share Company to be a holding company, part of the General Electric and a few other concerns. With the control fully secured of equipment and appliances, fuel and the output of generating companies, the only factor which might remain uncontrolled would be the labor element. There is no doubt that immense power corporations, acting under a uniform policy, dictated by the central agency holding control, will make a serious and well financed attempt to control the labor in the field of the electrical industry....
>
> We have gained a little, but it must be recognized and frankly admitted that the future power and success of the Brotherhood depends in no little measure on the way with which we meet and handle the public utility situation, and especially those in the power industry which are at this period undergoing the greatest change ever brought about in any industrial world.
> "President Noonan's Remarks," *IBEW Convention Proceedings 1925*

Above: Industrialist Samuel Insull's vast and highly leveraged utilities holding company collapsed in 1932, taking with it the life savings of some 600,000 investors. That same year, holding companies controlled over 78 percent of electrical generation. www.clipartlogo.com image.

Right: IBEW #77 dues book from July 1919, listing $69.60 in "strike donations." This was almost twice as much as the per capitas (standard union dues) collected that month. Local #77 Archive. Below: WWP linemen with truck. Avista Archive.

DAY BOOK—Receipts

(Enter daily Receipts from all sources received. This must be done regularly at each meeting.)

DATE		ass	p.cap.	RECEIPTS $	Cts
Rec. no July 1919	AMOUNT BROUGHT FORWARD, -	22.25	128.40	354	60
552282	Loyd Evans	50	1.20	3	50
83	W. C. Broughton	50	60	1	50
84	H. E. Hanlan	50	.60	2	00
85	F. B. Hoffman	50	1.80	5	00
86	J. D. Stewart		.60	1	00
87	H. R. Wiedenback (unit)		1.00	3	00
88	Ed. Brooks	50	1.20	2	50
89	R. D. Bush		1.80	4	50
90	Geo. H. Whipple	75	2.20	6	75
91	W. G. Skelton	75	2.20	5	65
92	Chas. Crass	50	.60	2	10
93	A. J. Cherrier	.50	1.20	2	50
94	Ralph Cook	50	60	1	50
95	R. H. Pond	50	60	2	10
96	Fay Ewer	75	2.60	5	55
97	Andy Gilbertson	50	.60	2	00
98	F. M. Bird		.60	1	50
99	Frank Burns		1.20	3	20
552300	M. A. Overackis (app)			1	00
1	F. E. Hayman		1.80	4	60
2	F. C. Barham		1.80	4	60
3	Bert Pinney	50	2.80	7	00
4	Chas. Walker	50	60	2	00
5	T. Jones		60	1	50
6	Geo. Johnston	50	1.20	3	50
7	Fred Allen	31.00	60	1	50

Strike donations

Rec. from - Frank McGovern $39.60
         Chas Lyons   15.30
         F. T. Rauen   15.00
               $69.80

Chas. Hartzell (for workers)   .25   25

             159.40 $ 416 50

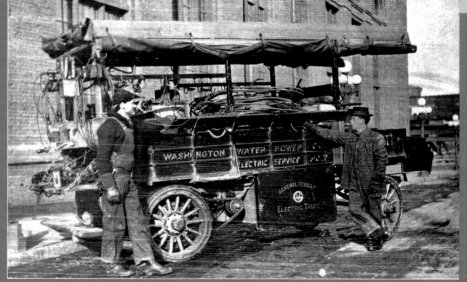

The issue of those who would come to be called "boomers" was also of concern to President Noonan. Groups of linemen were following utility construction work through the jurisdictions of individual IBEW locals without being members of these locals. There was dissension between locals about who within the IBEW should have control over this work. President Noonan described the realities of the contracting companies. These contractors would not accept IBEW's traditional jurisdictional model, using defined geographic locals:

> [T]he most strenuous objection that they [contracting employer companies] offer... are to have their men organized under a system that compels them to discharge all of the men they have trained in powerhouse construction at the completion of one job and hire an entirely new force in the locality in which their next operation is to be located. They claim to have built a force that they term construction men, expert in that line, and they state it is less expensive to meet the troubles occasioned by non-union men and the building trades' objection than to train a new force of men for each job.
>
> In the near past, some of these companies stated that if our laws were to allow some flexibility in the situation, allow them to maintain a construction crew of a reasonable amount of men, and to move to any locality and employ the others necessary for the job from whose jurisdiction the work is located (and this would be the great majority of the men employed), they would agree to operate as a union shop under an international agreement that would allow for their meeting the wage scale and working conditions of the various localities. They cited the fact that other international unions have such agreements with contractors doing interstate business, that the Bricklayers, Carpenters, Ironworkers and other unions have such agreements or arrangements with firms who do construction in all parts of the country, that allow them to move their experienced men so long as these men have their required standing in their international unions, and at least a sufficient force for directing the work and performing some of the most technical details.
>
> "President Noonan's Remarks," *IBEW Convention Proceedings 1925*

What the IBEW needed was "flexibility." In exchange, many of the construction companies were willing to agree to operate under an "international agreement" that laid out wage scale and working conditions "of the various localities." Noonan continued:

> An agreement of this kind would, of course, interfere with the working rules of some local unions which provide that all men must be members of that local union and in some instances, that the foremen must be members of that particular local for a specified period prior to their taking any job in a supervisory capacity.
>
> "President Noonan's Remarks," *IBEW Convention Proceedings 1925*

Research did not uncover any discussion of this proposal within Washington State IBEW locals until later; however the status and jurisdictions of IBEW locals in the state as of early 1925 is not well documented, especially that of #77.

The Giant Power Movement looked different to IBEW members in the Pacific Northwest. The IBEW had withdrawn #77's charter shortly after the Seattle General Strike of 1919 and had chartered #944 days later to replace it. During #944's existence (1919–1925), there are no existing minutes for #77. But Local #944's minutes include constant references to a group of IBEW members who continued to meet as #77, and who continued to refuse to recognize #944 as their IBEW local.

In December 1925, #944 voted "that the Charter Number would be changed to #77." The IO agreed, and #77 was rechartered with jurisdiction "over Seattle and vicinity." Hand-written below the signatures on this new charter was this additional description of its jurisdiction:

> [M]ixed together with workers coming under such classes and jurisdictions as defined and approved by the International office.
>
> *IBEW LU 77 charter document,* IBEW archives (IBEW International 2011)

The issue of which workers should be in which Washington State local would come to a head in the late 1930s, but the factors that contributed to intrastate jurisdictional conflict began shortly after electrical service came to the region.

Northwest linemen worked directly for utilities rather than for contractors before the Giant Power Movement and the rapid expansion of the electrical grid that came with it. As utility employees, they worked within that utility's service area (usually city limits), at its hydropower generation site, and on the transmission lines between the two.

Above: Washington State congressional delegation in Senator Homer T. Bone's office, Washington, D.C., February 1941. University of Washington Libraries, Special Collections, UW27134.
Right: Seattle Mayor Bertha Landes' billboard, late 1920s, promoting SCL as a publicly owned utility, unlike PSP&L. University of Washington Libraries, Special Collections, UW35559.

Washington State Creates PUDs

ALTHOUGH there is no documented discussion of "boomer" or "construction men" by Washington locals immediately after Noonan's 1925 convention proposal, there *was* major disgruntlement with privately owned utilities in Washington State. Electrical service included public transportation via electrical trolley systems, as well as residential, industrial, and in a few locations, agricultural electricity. Across the state, customers of the ever more consolidated electrical utility industry wanted greater uniformity in the quality and availability of their electrical service, and they wanted electricity at "fair rates."

Antipathy toward "big business" (a term that came into use in 1905) in general, and toward the utility industry in particular, brought sweeping popular interest in publicly *owned* power. (Remember the term "public power" refers to power provided to the public, not necessarily owned by the public.)

Farmers and those outside municipal service areas were increasingly frustrated at their lack of access to electricity, while ratepayers and trolley riders within many cities became equally dissatisfied with their private utility companies' high rates and variable levels of service.

Resentment toward large corporations and anger at the lack of access to affordable electricity generated changes in public opinion, political action, and eventually in state and federal law.

In the Pacific Northwest, the public's demand for better access to electricity found common ground with the federal government's interest in Giant Power, which sought to identify and develop a coordinated system of electricity, including a coordinated system of hydropower generation.

As part of a natural resources inventory mandated by the Rivers and Harbors Act of March 1925, the U.S. Army Corps of Engineers (USACE) completed a report on the Columbia River Basin. This study evaluated the Columbia's potential for irrigation, flood control, hydropower, and navigation.

For almost ten years prior to the USACE study, Washingtonians had been arguing over two approaches to damming the Columbia. The "pumpers" advocated a dam with pumps that would irrigate the Grand Coulee. The "ditchers" wanted to divert water from the Pend Oreille River and use a canal (or ditch) and gravity to irrigate. Neither group had financing.

Meanwhile, municipally owned utilities provided better service and cheaper rates than their privately owned competitors. In 1924 State Representative Homer T. Bone put forward legislation that would have allowed cities to provide electrical service to bordering entities. Private utilities countered with legislation that would have financially penalized any city extending infrastructure or service area beyond its corporate limits. Both proposals were on the fall 1924 general ballot. Both were defeated.

Supporters of public utility ownership regrouped. In 1929 their coalition, led by the Washington State Grange, ran a signature drive for Initiative #1, allowing rural communities to form their own publicly owned utilities. More than 60,000 signatures, more than twice the number required, were gathered, and the initiative moved to the Legislature.

Above: WPA work crew. University of Washington Libraries, Special Collections UW35571. (PSP&L photo). Right: National levels of all IBEW membership (inside/wiremen and outside/linemen), 1931–1936. IBEW IO Archive. Below: SCL Gorge Dam workers displaying fish that they caught, probably on Newhalem River, Washington. University of Washington Libraries, Special Collections UW27569.

RELATIVE EMPLOYMENT AND UNEMPLOYMENT OF I.B.E.W. MEMBERS

1931–1936

 EMPLOYMENT

 UNEMPLOYMENT

NOTE:—EACH I.B.E.W. SYMBOL REPRESENTS 10% OF FULL EMPLOYMENT

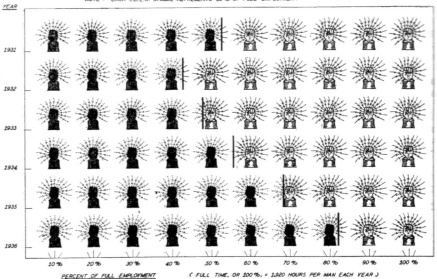

PERCENT OF FULL EMPLOYMENT

(FULL TIME, OR 100%, = 1,920 HOURS PER MAN EACH YEAR)

Black Tuesday and the Great Depression

THEN came "Black Tuesday"—October 29, 1929—the day the U.S. stock market crashed.

In Washington State the public power movement kept rolling even after Black Tuesday. When the Legislature took no action on Initiative #1, the measure came up as a statewide ballot measure in 1930. The AFL joined the Grange to get out the vote. The measure passed with 54 percent of the vote according to the Washington PUD Association.

The new PUD law, RCW 54, Public Utility Districts, went into effect in 1931.

> The purpose of this act is to authorize the establishment of public utility districts to conserve the water and power resources of the State of Washington for the benefit of the people thereof, and to supply public utility service, including water and electricity for all uses.
> *Title 54 RCW, Public Utility Districts*

From the beginning PUD law overrode other state law that might impede or limit its intent:

> Adjudication of invalidity of any section, clause or part of a section of this act shall not impair or otherwise affect the validity of the act as a whole or any other part thereof.
> The rule of strict construction shall have no application to this act, but the same shall be liberally construed, in order to carry out the purposes and objects for which this act is intended.
> When this act comes in conflict with any provision, limitation or restriction in any other law, this act shall govern and control.
> *Title 54 RCW, Public Utility Districts*

Some call Black Tuesday the beginning of the Great Depression. Many others, including John Kenneth Galbraith, saw it as the day Wall Street succumbed to the financial instability and economic hard times the general economy had been experiencing for some time.

Rampant financial speculation and leveraged investment within the utility industry and market were key contributors to national economic instability and the stock market crash that followed.

> By 1929, sixteen groups controlled 92 percent of the nation's electrical power output. During the first decade of the FPC's [Federal Power Commission's] existence, the electric utility industries had undergone tremendous growth, far outpacing both state and federal regulation. Approximately 15 percent of the electricity generated in the United States moved across state lines, and the percentage was steadily increasing.
> Cantelon, "The Regulatory Dilemma of the Federal Power Commission, 1920–1977," p. 67

The Federal Trade Commission (FTC) began a formal investigation of holding companies in 1928. Historian Howard Zinn summarizes what many experienced during the Roaring Twenties:

> [P]rosperity was concentrated at the top. While from 1922 to 1929 real wages in manufacturing went up per capita 1.4 percent a year, the holders of common stocks gained 16.4 percent a year. Six million families (42 percent of the total) made less than $1,000 per year. One-tenth of one percent of the families at the top received as much income as 42 percent of the families at the bottom, according to a report of the Brookings Institution.
> Zinn, *A People's History of the United States*, p. 382

By 1933 workers in the United States had been pummeled by the Depression, with unemployment at 35 percent. Several of #77's retired linemen recounted growing up in extreme poverty during this period. Several described getting their first union job as "dying and going to heaven."

The federal government, under President Herbert Hoover, started a number of programs designed to spur economic recovery, without positive results. Meanwhile, the utility industry continued to do far better than most other economic sectors.

> At the height of the Depression, in 1933, [power industry] revenues were off only 11 percent. Gross agricultural income by comparison had fallen by 58 percent in three years, and manufacturing income declined 46 percent. Total power production dropped by about one third between 1929 and 1935. Much of the decline in industrial power usage was made up by increased sales to the 20 million households, which eventually provided one third of the electric industry's revenues.
> Rudolf and Ridley, *Power Struggle,* p. 62

In 1930, the Federal Power Commission (FPC) had been retooled, and began to survey power production, availability, and rates across the United States. Industrialist Samuel Insull's vast and highly leveraged utilities holding company collapsed in 1932, taking with it the life savings of some 600,000 investors. That same year, holding companies controlled over 78 percent of electrical generation.

Franklin D. Roosevelt defeated Herbert Hoover, and took office in March 1933. He immediately began to implement the New Deal, which was designed to provide immediate economic relief. FDR sent Congress bill after bill. One of these, the National Industrial Recovery Act (NIRA), established the National Recovery Administration (NRA). The NRA created the Public Works Administration and granted workers the right to organize. The NRA also authorized trade associations to draft industrial codes provided they included maximum hours and minimum wages.

Labor unions, including the IBEW, expended their resources to attend endless NIRA hearings, while industry leaders, in no hurry to commit to binding hours or wages, instead moved quickly to create company unions so their workers were "organized."

FDR took immediate additional action toward more equitable access and charges for electricity:

> Roosevelt and others... envisioned power flowing to the rural areas, particularly farms... provided at the same cost as in urban areas.... everyone would pay a slight premium so that all could enjoy a single rate—a "postage stamp" rate. The concept is borrowed from one-price postage stamps, which will deliver a letter across the street or across the nation.
> Northwest Power and Conservation Council website, "Bonneville Power Administration, History"

The Tennessee Valley Authority (TVA) was authorized by Congress two months after FDR was in office,

> to address a wide range of environmental, economic, and technological issues, including the delivery of low-cost electricity and the management of natural resources.
> Tennessee Valley Authority website, "From the New Deal to a New Century"

In Washington State, FDR followed through on his campaign promise to develop the Columbia River. His hope was to create something like the TVA for many major river systems in the United States, including a "Columbia Valley Authority." FDR backed the idea of a Grand Coulee Dam but not federal funding on the entire project. Instead, he took action to secure federal funds as seed money.

> When Congress balked at Grand Coulee Dam over whether it was needed and its cost, Roosevelt used his authority under the National Recovery Act to provide the initial appropriation.
> Northwest Power and Conservation Council website, "Bonneville Power Administration, history"

Washington Governor Clarence Martin set up a commission to secure funding for construction and oversight of the project. The U.S. Bureau of Reclamation (USBR) signed on to "prepare plans, conduct surveys, and begin preliminary work on the project," and ground was broken in July 1933.

Days after he authorized limited funding for the Grand Coulee Dam, FDR authorized about nine times as much for construction of the Bonneville Dam, under the U.S. Army Corps of Engineers.

Car and houses beside SCL 165 KV transmission line and bus structure, North Substation, 1924.
University of Washington Libraries, Special Collections UW27567.

Upper left: East Coast investors (or speculators) milk the power trust that feeds upon Washington's natural resources. From the Grange News (Seattle) September 5, 1930. Rick Luiten Archive.

Above: Farmer and contractor shake hands to commemorate the farm getting electrical service, 1929. IBEW IO Archive.

Left: REA projects in the Northwest. Avista Archive.

Federal Law:
The Public Utility Holding Act

IN 1934, a year after both of the state's huge federal dams were underway, Washington's first PUDs were created. In 1934 Mason County PUD #1 and #3 were established, Mason County PUD #1 went into operation, and Benton and Franklin County voters approved the first countywide PUDs.

Congress took a different approach toward encouraging rural electrification. It authorized a ten-year loan plan for existing utilities that chose to extend their services into rural areas. There were few takers due to the relatively high costs and low profits of such extensions.

Using his authority under the Emergency Relief Appropriation Act of 1935, President Roosevelt created the Rural Electrification Administration (REA) in May 1935 (Executive Order 7037).

> At the time, only 10 percent of the nation's farms had electricity. The situation was much better in the Northwest, where 47.5 percent of the farms in Washington, 27.5 percent of the farms in Oregon, and 29.8 percent in Idaho had electricity. In Montana, only 5.5 percent of the farms had electricity.
> Northwest Power and Conservation Council, "Rural Electrification"

The REA was immediately popular but it had difficulty hiring enough skilled workers (engineers and linemen, for example) when it could only pay wages at rates authorized for "unemployment relief."

The federal Public Utility Holding Company Act (also known as the Wheeler-Rayburn Act, or PUHCA) also passed in 1935. This sweeping legislation forced a restructuring of the utility industry. PUHCA required a utility holding company to either limit its activities to a single state, so that state could effectively regulate its activity, or to obtain prior approval from the Securities and Exchange Commission (SEC) before conducting *any* non-utility business.

> Because the SEC strictly enforced the divestiture provision of PUHCA in its proceedings and ordered divestiture of all corporate holdings except for a single integrated electric system, the affected holding companies filed voluntary divesture [sic divestiture] plans. As a result, by 1948 holding companies had voluntarily divested themselves of assets worth approximately $12 billion and the number of subsidiaries controlled by affected holding companies was reduced from 1,983 to 303.
> Wikipedia, "Public Utility Holding Company Act of 1935"

The following year, 1936, the REA gained administrative autonomy under the Rural Electrification Act. (This is a bit confusing because the agency, the administrative act, and individual Rural Electric Associations share the same acronym.)

The Rural Electrification Act of 1936 (Act) extended the benefits of electric power to the nation's farmers through a low-cost loan program. Before that Act was passed, only a handful of cooperatives existed; six years later, over 800 had been formed.
Public Power Council website, "Public Power History"

In Washington State, the State Supreme Court upheld the PUD law (RCW 54) while the Washington State Grange worked with PUD supporters to get public utility districts onto as many local ballots as possible. Supporters also sponsored a meeting of PUD commissioners who then formed a trade organization, the Washington Public Utility Commissioners Association (WPUCA, now the Washington Public Utility Districts Association, or WPUDA). By 1944, more than twenty-four counties had created PUDs although few of them were in operation. Once created, PUDs began to take ownership of privately held hydroelectric sites and facilities through condemnation.

The burgeoning development of Northwest hydropower at federal, private, PUD, and cooperatively owned sites drove the federal creation of the Bonneville Power Administration in 1937 (under the Bonneville Power Act). FDR still hoped for a Columbia Valley Authority that would function like the TVA. The BPA was conceived as a temporary stand-in until such an authority could be created. As such, the BPA was directed to govern the marketing of electricity from the federally owned Grand Coulee and Bonneville dams while creating a regionally integrated transmission system.

One of the BPA's dictates would shape the future of the Northwest utility industry: its mandated preference to sell its electricity to publicly *owned* utilities.

In order to ensure that the facilities for the generation of electric energy at the Bonneville project shall be operated for the benefit of the general public, and particularly of domestic and rural consumers, the administrator shall at all times, in disposing of electric energy generated at said project, give preference and priority to public bodies and cooperatives...
U. S. Department of the Interior, The Bonneville Power Act

J.D. Ross was the BPA's first administrator. (He concurrently continued as superintendent of Seattle City Light, although he did so for free.) Ross pushed rapid expansion of the BPA's transmission system, carrying forward FDR's "postage stamp" rate concept, wherein all utilities paid the same rate for BPA-delivered power, no matter how far BPA lines ran to deliver electricity. Ross also sought out industrial customers to build the BPA's revenue base while encouraging more PUDs to come on line.

By 1940, thanks largely to the establishment of the REA, the completion of Bonneville Dam in 1938, and the creation of rural electric utilities, 59 percent of the farms in Oregon had electricity; in Washington, the figure was 71 percent.
Rural Electrification Administration, *A Brief History of the Rural Electric and Telephone Programs*, p. A-1

(This was a 32 percent increase in Oregon, and a 23.5 percent increase in Washington State, from 1935.)

As federal and state government promoted rapid expansion of the electrical utility system, huge changes in federal labor law were also passed. FDR signed the National Labor Relations Act (NLRA, also known as the Wagner Act) in July 1935. The Wagner Act made explicit the unions' right to organize and the employers' obligation to bargain with unions on wages, hours, and other terms and conditions of employment. This act passed on the heels of a Supreme Court decision that found the National Industrial Recovery Act (NIRA) unconstitutional.

In the two years after NIRA's passage, the IBEW put much of its resources into NIRA hearings on the electrical industry. The IBEW supported the AFL in seeking to keep control of journeyman craft work rather than opening its doors to non-journeymen. In the case of the IBEW this meant keeping out electrical workers with less craft training, be they wiremen, linemen, or manufacturing workers.

In *Dreams of Dignity,* Palladino makes the point that many IBEW members knew how difficult it was to keep their worksites unionized and felt a small journeyman-oriented membership guaranteed good work and good wages to those who were already in the IBEW.

Once the Wagner Act passed, the IBEW attempted to bring in non-journeymen and manufacturing workers by restructuring its membership, and expanding its "B" membership. B membership was a second tier of membership that did not include full voting rights, pension benefits, or insurance benefits. It was originally created to allow lower-paid telephone workers (largely female) to pay lower dues. The B membership was sometimes organized into sub-locals or independent locals rather than into existing electrical workers' locals.

Many in the IBEW were concerned that the influx of new B-members would eventually change the union's balance of power when B-members demanded A-membership. Large numbers of manufacturing workers would then become the controlling majority of IBEW, making jurisdictional decisions and directing the IO to focus on improving *their* working conditions and wages at the expense of journeymen wages and working conditions.

The IBEW therefore supported the AFL's work-around, which was to create separate "federal unions" for less-skilled manufacturing workers. (A federal union is chartered directly to the AFL without an international, and is often assigned to an already existing AFL international at a later date.)

To many workers who had never been in a union, this looked like an arrogant attempt to keep them from getting good work. The Committee for Industrial Organization (CIO), led by John Lewis of the United Mine Workers, formed within the AFL in 1935. A year later, in 1936, the AFL suspended the ten unions in the CIO.

That same year, the United Electrical Radio and Machine Workers of America (UE) was created with the purpose of uniting "all workers in our industry on an industrial basis... regardless of craft, age, sex, nationality, or creed, or political beliefs."

The UE used an aggressive strike policy to challenge the IBEW's jurisdiction over radio and manufacturing work, primarily on the East Coast. The AFL refused to affiliate the UE, so the UE continued as an autonomous union and grew rapidly.

In 1937 the UE expanded its focus and began to organize utility workers as well as those in radio and manufacturing. The IBEW saw utility work as a key part of its jurisdiction, but had a hard time organizing it in the East and parts of the Midwest. IBEW President Daniel Tracy, a former lineman, had gotten the AFL to grant the IBEW exclusive jurisdiction over *all* utility workers in 1933. Tracy wanted to bring production and maintenance workers in, along with the utility linemen.

Just before the Wagner Act passed in 1935, the IBEW had attempted a major organizing campaign for utility workers in the Midwest, which was "sometimes violent," according to Palladino. Immediately after passage of the Wagner Act, Tracy targeted federally funded public utility projects, starting with the TVA. The IBEW was a signatory to TVA's "seven-local labor management council" and had some working agreement by 1937, according to a Tracy memo cited by Palladino.

The UE gave the IBEW a good taste of how the Wagner Act had changed the labor landscape. Although it provided legal leverage for union activity, the Wagner Act also opened the door to inter-union competition, and the UE made huge gains in what had been the IBEW's jurisdiction.

In 1938 the Committee for Industrial Organization became its own labor federation, the *Congress* of Industrial Organizations (also CIO). One of the first actions of the CIO was to charter the UE. By 1940 the UE represented more than 600,000 workers.

The UE did not have much of a presence in Washington State. Union linemen in the Northwest were IBEW members, but did not receive a great deal of support from the IO during this period. In addition, there were no IBEW conventions from 1929 to 1941 that might have allowed the membership to meet and direct the International. There were, however, some regional challenges to IBEW representation from CIO affiliates and/or IWW members. Historical records show few references to specific CIO or IWW organizing campaigns relating to the IBEW in Washington State, although legend has it there were significant challenges in the Spokane area. The bigger jurisdictional fight seems to have been between IBEW locals.

The tremendous expansion of electrical generation, transmission, and delivery within Washington State created new jobs. This utility work ranged across the entire state, far from the traditional IBEW turf of municipally based locals. There was work on new utilities (PUDs and REAs), work to expand existing utility service areas, work on the federal dams and BPA transmission lines, and work maintaining and operating all of these utilities.

Local #77 represented "Seattle and vicinity" going into the Depression. In the early 1930s, hard times seem to have fostered solidarity, at least in Seattle. Locals #77 and #46 (Seattle's inside wiremen's local) actually co-wrote their November 1933 article in *The Electrical Worker*. The article ran under the header "Harmony in Seattle."

> Some members of our organization [the IBEW] [would] be astonished to know that the inside and the outside electricians in Seattle are working in harmony. We are holding Joint executive board meetings and working on plans to prevent any friction that may tend to develop in the future.

(...pert, Class II)	*109.00*	*112.50*
Storeroom		
Sr. Warehouseman	145.00	148.50
Warehouseman, Seattle	124.00	127.00
Warehouseman	109.00	111.50
Janitors		
Senior Janitor	107.00	109.50
Janitor (Steam License)	97.50	100.00
Janitor	93.00	95.50
Elevator Operator	93.00	95.50
Drayage		
Truck Driver (Trailer)	155.50	159.50
Truck Driver	146.00	149.50
Apprentice Linemen (Seattle)	**(per hr.)**	**(per hr.)**
1st 6 months	0.7375	0.755
2nd 6 months	0.7825	0.8025
3rd 6 months	0.8275	0.8475
4th 6 months	0.8725	0.8950
5th 6 months	0.9175	0.9400
6th 6 months (third year)	0.9625	0.9875
Apprentice Linemen		
1st 6 months	0.6500	0.6675
2nd 6 months	0.6975	0.7150
3rd 6 months	0.7450	0.7650
4th 6 months	0.7925	0.8150
5th 6 months	0.8400	0.8625
6th 6 months (third year)	0.8875	0.9100
Apprentice Wiremen (Seattle)		
1st 6 months	0.7875	0.8075
2nd 6 months	0.8225	0.8425
3rd 6 months	0.8575	0.8800
4th 6 months	0.8925	0.9150
5th 6 months	0.9275	0.9525
6th 6 months (third year)	0.9625	0.9875
Apprentice Cable Splicers (Seattle)		
1st 6 months	0.7875	0.8075
2nd 6 months	0.8375	0.8600
3rd 6 months	0.8900	0.9125
4th 6 months	0.9400	0.9650
5th 6 months	0.9925	1.0175
6th 6 months (third year)	1.0425	1.0700

L. W. STOCKWELL

1936

R AGREEMENT

BETWEEN

Puget Sound Power & Light Company

AND

International Brotherhood of Electrical Workers

LOCAL No. 77

7/15/36 – 7/15/37

Clockwise from upper left: Wage schedule from the first agreement between PSP&L and IBEW, 1936–1937, IBEW Local No. 77 Collection, University of Washington Libraries, Special Collections, UW35593. Cover of the first agreement between PSP&L and IBEW, 1936–1937. IBEW Local No. 77 Collection, University of Washington Libraries, Special Collections, UW35593. Dismantling 22Kv. Tower at WWP's Snake River Crossing. Avista Archive. WWP linemen at work. Avista Archive.

Dismantling 22 Kv. Tower – Snake River Crossing – 1931

President William Grace, of Local No. 46, and President C. L. Hardy, of Local No. 77, have requested that this letter express the desire of the officers of each local union that our members try not to be too touchy about jurisdiction rights and membership in a certain local. This is not meant to be a polite request that the other fellow keep off our toes, but a suggestion that the members wear toe protectors when they come down to the Labor Temple.
The Electrical Worker, November 1933

But when more work came along, and with it the possibility of larger jurisdictions, IBEW locals in the Northwest began competing for the same workers.

In November 1933, Portland's #125 preemptively claimed whatever electrical work might come from the proposed Bonneville Dam and warned off IBEW travellers:

Regardless of newspaper publicity, chamber of commerce and real estate advertising, we would ask all members to keep away from Portland, or at least not to migrate here with the idea of securing work on the Bonneville Dam.
The Electrical Worker, November 1933

Seven months later, #77 expressed its own ideas about at least some of the same work, after summarizing organizing plans in the Seattle area:

The big men like Frank McLaughlin, president of the Puget Sound Power and Light Company and Dr. J.D. Ross, superintendent of Seattle's 1,000,000-kilowatt light plant, tell us to go ahead and organize.

We plan to have both companies organized 100 percent within the near future. Then we will be in a position to aid the government in handling the Coulee and the Bonneville jobs on the Columbia River.
The Electrical Worker, June 1934

Local #73, out of Spokane, moved to represent workers at Grand Coulee, and had a sub-local, also referred to as Unit #1, at Grand Coulee that held its first unit meeting in August 1937.

Local #77 continued to expand its jurisdiction and membership. At the beginning of 1936, two-year-old Centralia-Chehalis #745 was amalgamated into #77. That same year the IBEW approved #77's revised bylaws, which gave #77 jurisdiction "over all outside electrical work and outside electrical workers in Seattle, Washington, and vicinity." In June 1936, #77 was proud to report:

Through the leadership of our international representative, Scott Milne, and our business manager, George Mulkey, Local Union No. 77 has grown in three years from a membership of 125 to over 1,200. We will have practically a closed shop in the Pacific Northwest, and therefore can view our problems from our employers' viewpoint. Their problems are our problems; they need a larger market for electric power, and we need more jobs.
The Electrical Worker, June 1936

A year later, in May 1937, #77's charter was amended "to include B-type membership."

This decision seems to have had as much to do with the IO's membership structure in relation to its pension benefits as it did with changes in #77's membership. Local #77's explicit addition of B-members may also have been prompted by the formation of #741, a Seattle local chartered less than a year before. Local #77 Business Manager George Mulkey assisted #741, a "radiomen's local," in organizing the "hotel telephone girls" into #B-741. Together #741 and #77 secured "a closed shop agreement and from 25 to 35 percent increase in wages and shorter hours," according to Irving Pattee's report in the May 1937 *Electrical Worker*.

Later that summer, #741 was amalgamated into #77, bringing along #741's Business Manager Mullaney and the "hotel telephone girls" of #B-741, also known as a Private Branch Exchange (PBX) unit.

In August 1937, #77 went public with its plan to organize *all* of Washington State's electrical workers with a page-and-a-half spread in *The Electrical Worker*, "Local Union Organized on State Wide Basis."

Left: Electric choppers, mash, and emery wheel for poultry farm use, in a promotional photo by PSP&L. University of Washington Libraries, Special Collections, UW35573.

Above: Unknown #77 member at work. Note his ladder is roped to "bells." Local #77 Archive.

Below: Local #77 members in shop. Local #77 Archive.

Local #77 "Goes Where Electricity Goes"

MANY IBEW locals were fending off other AFL unions as electrical work expanded. Ironworkers and linemen squared off over who should have jurisdiction over constructing steel towers. In December 1937, #77 used its monthly report to respond to this and other disputes between AFL unions:

> For Local No. B-18's [Los Angeles #B-17's article in an earlier issue of *The Electrical Worker*] don't get high hat over the fact that the structural iron workers are trying to get the towers away from you; they even had us declared unfair to (WHAT?) organized labor by the building trades in half the towns in Washington because we claimed jurisdiction over bolted towers and even went so far as to attempt to picket one job of switch racks.
>
> The office workers objected to our taking in the office help of the P.S.P.& L. Co. and the technical engineers are in a good place with the City Light to swing in on us as foremen of line gangs. Maybe Papa Green [AFL President William Green] and the rest of the big leaders of the A. F. of L. had better remember that at a certain convention not so long past the electrical workmen were awarded a few things, and, too, maybe this C.I.O. trouble can be classified as an ill wind that may blow some good, if it only clears up this question of when an industry can best be organized industrial or by craft.
>
> *The Electrical Worker,* December 1937

At some point prior to World War II, the majority of #77's members were no longer linemen. Exactly when this shift occurred is unclear, but it seems to have come from #77's two-fold expansion during the late 1930s: #77 was reaching far beyond "Seattle and vicinity," and the local wanted to represent all the workers at a utility, no matter what their work.

Local #77 used IBEW President Tracy's TVA model to represent not only utility linemen, but also utility employees in generation, operations, maintenance, office work, groundskeeping, and other work. Whoever worked for a utility was welcome in #77, even a gravedigger, according to legend. At the same time #77 was organizing and representing radio repairmen, telephone workers, and other non-utility workers.

The actions taken by #77's organizers and members to secure what eventually became its current jurisdiction are shrouded in a prolonged and poorly documented period of expansion. Given all the turf wars, more properly referred to as jurisdictional disputes—disputes with other AFL craft unions, disputes with other IBEW locals, disputes with the CIO, and disputes with non-union utilities and contractors—how *did* #77 gain jurisdiction over such a large geographic area and such a wide variety of work?

The answer seems to be: heroic organizing by #77's staff and linemen, support from the IO, and a habit of taking on one employer at a time.

Local #77 business manager George A. Mulkey was one in a million. He came up as a lineman, and he worked with the linemen in #77 on every campaign. He seems to have been everywhere, from Olympia to Spokane to Mason County, and he had a way of efficiently "softening" employers until they signed, using the threat of strikes and other hard-ball tactics. At what point #77 began to have actual units is unclear. It does not appear it had any sub-locals other than when B-77 represented only B-members.

By inference it seems the IO sometimes supported the creation of sub-locals (or units) by existing IBEW locals and sometimes supported the creation of new, autonomous IBEW locals. At least in Washington State, the sub-local model seems to have been applied to utility work, whereas the autonomous locals seem to have been used when organizing B-membership worksites. Several B-locals were amalgamated into older IBEW locals.

Local #77's main jurisdictional competition seems to have been IBEW #73. According to the February 1938 *Electrical Worker*, "Sub-Local No. 73. Grand Coulee District" was created November 1, 1934, and was still a sub-local of #73 in 1938. Both locals also mention organizing and/or representing Central and Eastern Washington rural electric associations (REAs) and rural electric cooperatives (RECs, also known as coops) during the mid- to late 1930s, but make little mention of each other.

Across the Northwest, IBEW locals, including #77, organized the new publicly owned utilities as they were created. In 1940, the Northwest Public Power Association was created as an association of REAs and publicly owned electric utilities, with members in Alaska, Washington, Oregon, Idaho, and Montana.

By the end of 1940 there were twenty-two PUDs in Washington State, although only the Skamania PUD was actually operating at the time, according to the Washington Public Utility Districts Association (WPUDA).

It was during this critical period that #77 gained a reputation for full worksite representation that has become its tag line: "#77 is not a 'country club' union." (This expression is used to describe the difference between a traditional craft union and #77, in that #77 represents all utility workers, not just craft journeymen.)

At any rate, between 1937 and World War II, #77 did "follow electricity" to organize Washington State's utility work across all job classifications. The local focused on organizing entire workplaces and employers. From the beginning, #77 pegged all its members' wages to those of the linemen. In addition, #77 was willing to represent workers at a utility after construction was complete, while also representing boomer linemen who followed construction work.

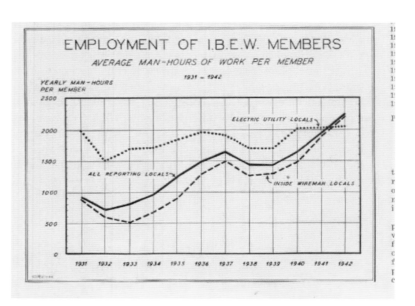

Left: WWP linemen, using leather gloves with rubber liners. Avista Archive.
Below: Employment levels of IBEW members 1939–1942, comparing utility work to inside or wiremen's work. IBEW IO Archive.

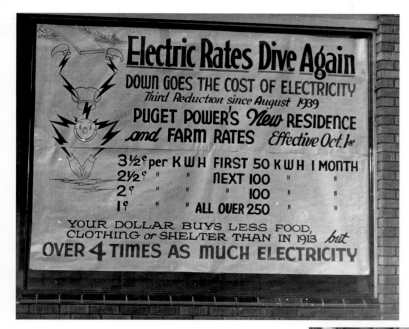

Left: A 1939 poster encouraging consumers to use more electricity. University of Washington Libraries, Special Collections, UW35570.

Below left: Mr. and Mrs. A.E. James, at Tekoa, December 1938. The photo is labeled, "Mr. James is a Water Power Veteran of some 25 or 30 years, a Substation man, transformer expert." Avista Archive.

Below right: PSP&L crew running underground cable down a manhole. University of Washington Libraries, Special Collections, UW35562.

Left: PSP&L crew running underground cable into a trench. University of Washington Libraries, Special Collections, UW35563.

Right: WWP crew with roped mountain goat, January 1938. Avista Archive.

PSP&L World War II support poster. University of Washington Libraries, Special Collections, UW35575.

Heading into War

AT the national level, the IBEW seems to have had its hands full trying to deal with the exploding need for factory workers as World War II broke out in Europe. The union movement in the United States knew its members' work and wages would be affected by war, and not necessarily for the better. Many unions, especially those in the AFL, wanted to avoid repeating both the witch-hunts for leftists and the decline in worker prosperity that came during and after World War I.

In July 1940, IBEW President Tracy was appointed Assistant Secretary of Labor, "in charge of military and war-related construction," according to Palladino. With Tracy's appointment the IBEW felt it was in as good a position as it could be relative to federal labor policy, and seems to have focused on making sure it had jurisdiction over as many aspects of defense and war-related work as possible. The August 1940 issue of *The Electrical Worker* was entitled "Total Defense" and ran pages of information about how to register through the federal government for jobs "fashioning defense war weapons... [and] mobilizing skills in all fields."

In an editorial in the same issue, the IBEW used growing political attacks on union "Bolshevism" to attack the UE as Communist:

> The United Electrical, Radio and Machine Workers of America is everywhere regarded in the labor movement as a communist-led and communist-controlled organization. James Carey, the young president, declares that he is not a communist and he is often exonerated of this stigma by friends, but it is also true that Carey takes his orders from the communist leaders of the executive council.
> *The Electrical Worker,* August 1940

The editorial also denounced UE charges against Tracy and the IBEW, saying they served to "blacken the character of an honest man and also to belittle the judgment of the President of the United States."

As war-related work expanded, the IBEW also moved to secure electrical work for its members by creating a system of certification for those going into construction. In September 1941, the IBEW and NECA (National Electrical Contractors Association) agreed to nationwide standards for apprenticeships by signing the National Apprenticeship Standards for the Electrical Construction Industry. This program was designed to cover inside electrical work by electrical contractors.

Then came Pearl Harbor, and the U.S. was at war.

During World War II, the IBEW more than doubled its membership, to over 360,000, as shipyard and factory work to build war machinery and ammunitions were all fast-tracked. The IBEW reports that something like 10 percent of its membership joined the armed forces. (Local #77 and #46 members who were in the military during World War II were able to come back into these locals under favorable terms after the war.)

Left: Female meter tester at PSP&L, 1944. WWII brought women into traditionally male positions in many industries, at least temporarily. University of Washington Libraries, Special Collections, UW35560.

Below: Journeyman teaching apprentice to oil threads. Avista Archive.

Opposite: 1941 PSP&L billboard and posters in Bellingham. University of Washington Libraries, Special Collections, UW3555.

During the war, the IBEW and other unions were regulated by a number of federal laws and labor-related boards, such as the War Production Board, the National War Labor Board, and the War Manpower Commission. In addition, days after Pearl Harbor, the AFL agreed to a no-strike policy, in part to put pressure on the CIO. President Roosevelt had been promoting a no-strike policy since before the U.S. entered the war. The AFL's position was significant enough that worldwide media took note:

> The executive council of the American Federation of Labor to-day unanimously adopted a no-strike policy for the duration of the war and decided to request President Roosevelt to create an agency similar to the Labor Board of the last war to provide a voluntary mediation and arbitration unit to handle all disputes without stopping work....
>
> [T]he AFL issued the following declaration. "We sincerely regret the destructive rivalry of labor organizations in past years and appeal to the Congress of Industrial Organizations to heal the labor movement's breach for the sake of common defense against mortal danger."
> *The Advertiser* (Adelaide, Australia), December 18, 1941

In the Pacific Northwest, the region's transmission system rapidly expanded to meet new industrial and military demands for power.

> At the outbreak of World War II, the cheap power generated at Pacific Northwest dams contributed greatly to the defense industry, particularly in the production of aluminum, as well as plutonium production at the Hanford Reservation [Hanford Site].
> Public Power Council website, "Public Power History"

Local #77 pushed to keep its jurisdiction and representation on war-related utility construction, as well as within various wartime federal projects.

According to Palladino, Secretary of War Robert Patterson worked with the IO in recruiting electrical workers for all of the Manhattan Project sites, including Hanford Engineer Works (HEW, now known as the Hanford Site). Anecdotally, members recall that #77 represented some electrical work at Hanford from its inception, but, not surprisingly for a top-secret military site, documentation is minimal.

Above: PSP&L linemen take a boat to check out damage from the Renton-Kent flood of 1946. University of Washington Libraries, Special Collections, UW35574.

Below: Local #77 members celebrate in the #77 Seattle offices, then on Melrose Avenue. Local #77 Archive, gift of Panhandler Peterson.

Spokane, Washington
January 2, 1952

HARRINGTON WIMBERLY, CHAIRMAN
FEDERAL POWER COMMISSION
WASHINGTON, D.C.

        AMERICAN POWER AND LIGHT COMPANY HAS AGAIN FILED A
NOTICE WITH THE SECURITIES AND EXCHANGE COMMISSION THAT IT INTENDS
TO DISPOSE OF THE WASHINGTON WATER POWER COMPANY TO PUBLIC UTILITY
DISTRICTS IN THE STATE OF WASHINGTON AND TO A NON-PROFIT CORPORATION.
LAST MARCH WHEN A SIMILAR ATTEMPT WAS MADE BY AMERICAN YOUR COMMISSION
INDICATED THAT SUCH A SCHEME MIGHT INVOLVE ACTION BY YOUR COMMISSION
UNDER THE FEDERAL POWER ACT. THE CITY OF SPOKANE IS TREMENDOUSLY
INTERESTED IN THIS PROPOSAL AND DOES NOT WANT ANYTHING DONE THAT MAY
JEOPARDIZE THE COMPLETION OF THE CABINET GORGE PROJECT OF THE WASHINGTON
WATER POWER COMPANY OR ANYTHING THAT WILL AFFECT THE CRITICAL POWER
SUPPLY AND THE WAR INDUSTRIES IN THIS AREA. I WOULD APPRECIATE IT VERY
MUCH IF YOU WILL ADVISE ME WHETHER YOUR COMMISSION IS INFORMED OF THE
PROPOSED TRANSACTION AND WHETHER YOU PLAN TO TAKE ANY STEPS IN REGARD
TO THE MATTER. SINCERELY BELIEVE THAT THIS WHOLE MATTER SHOULD BE
THOROUGHLY INVESTIGATED BY YOUR COMMISSION.

Left: January 2, 1952 telegram from Willard Taft (Spokane's acting mayor and commissioner of public utilities) to the FCC regarding APL's notice that it will "dispose of the Washington Water Power Company to Public Utility Districts in the State of Washington and to a non-profit corporation." Taft states, "The City of Spokane is tremendously interested in this proposal and does not want anything done that may jeopardize the completion of the Cabinet Gorge Project." Avista Archive.

Post-War Boom

ONCE the war was over, #77 and the entire IBEW regrouped. In 1946 plans were made for the first IBEW convention since 1941. Internal dissension helped fuel continuing jurisdictional challenges between the IBEW and the UE in parallel to AFL and CIO conflicts in the larger labor community.

Palladino identifies what she considers the key conflict within IBEW as the 1946 convention approached:

> [C]raft and industrial locals existed side-by-side in the IBEW by 1946, and jurisdictional tensions between the two shaped this [1946 convention] conflict even if they did not determine it.
> Palladino, p. 199

This craft vs. industrial conflict is particularly interesting relative to #77's successful industrial model and its expanding jurisdiction from the late 1930s into the 1950s.

Anticipating a contentious convention, IBEW President Brown tried to postpone it until 1947. This only solidified his opposition, including some members of the International Executive Committee (IEC). Former IBEW President Tracy's candidacy was orchestrated by two District vice presidents, West Coast District 9 VP Scott Milne and Midwest District 6 VP Mike Boyle, out of Chicago. (Tracy had resigned from his federal position thinking President Brown would then give him a leadership position within the IBEW, but Brown never made an offer.)

Tracy created some enemies of his own by violating the long-standing practice that the IO support an incumbent IBEW president, although Tracy was no longer with the IBEW. But when Brown fired "disloyal" vice presidents just before heading to San Francisco for the convention, he galvanized support for Tracy. As a record number of IBEW delegates converged, the local San Francisco press featured inflammatory coverage and predicted violence.

When President Brown opposed a constitutional amendment from Oakland #594 that would grant each District the power to elect its own vice presidents, he lost even more delegate support. Tracy won the election by less than 1 percent. Tracy supporters then won all but two of the District vice presidencies.

By the end of this 1946 convention, a new category of membership was created, the "BA" membership. BA members would pay full per capitas and receive full voting rights, but would not have any pension or death benefits, nor could they vote on these matters.

In the three months after the 1946 convention, the IEC took on redistricting. In early 1947, the number of districts increased from ten to twelve and a summary of the IEC's "special meeting" on restructuring was featured in the March 1947 *Electrical Worker:*

> Eleven resolutions on the subject [proposal for redistricting], mailed to the International Office since the Convention, were also studied and considered. These resolutions were all worded the same and originated in one local union. They were received from the following locals:
>
> 46 Seattle, Washington
> 73 Spokane, Washington
> 265 Lincoln, Nebraska
> 619 Hot Springs, Arkansas
> 574 Bremerton, Washington
> 12 Pueblo, Colorado
> 124 Kansas City, Missouri
> 460 Midland, Texas
> 991 Corning, New York

Note that there are no Northwest linemen's locals listed as petitioners. The article went on to explain specific changes to the West Coast (9th) District.

> After adding Alaska to this district and taking Arizona from it, the district would then consist of Nevada, California, Oregon, Washington, Alaska and the Pacific Islands.
> *The Electrical Worker,* March 1947

The full membership approved this restructuring proposal. Shortly after, one of the few remaining Brown supporters on the IEC, Secretary G.M. "Gus" Bugniazet, was forced to resign after twenty-three years of service.

The IBEW had barely finished its reorganization when Congress passed the Taft-Hartley Act (Labor Management Relations Act), over President Truman's veto, in June 1947.

Taft-Hartley made administrative changes to federal labor law including adding two positions to the National Relations Board (NLRB); changing the Federal Mediation and Conciliation Service (FMCS) to an independent agency (it had been under the Department of Labor); and establishing the option of an eighty-day injunction against striking, a "cooling-off period" to "protect the welfare of the nation."

The Taft-Hartley Act also outlawed secondary boycotts and closed shops, allowed states to enact right-to-work laws, required unions to file financial reports to the Labor Department, and required labor leaders to swear under oath that they were not Communists (the Supreme Court in 1965 ruled this last provision was unconstitutional).

The IBEW had the same reaction to the Taft-Hartley Act as many other unions:

> The Executive Council condemns the Taft-Hartley Law. One-third of our population have been stabbed in the back. Union members have not yet awakened to this fact. So your Council issues this statement: The law was written by the die-hard enemies of Unions. It is loaded with hate and bitterness. It puts class war into law. It is a hodgepodge, jigsaw law—filled with sneak wording and hidden meanings.
> *The Electrical Worker*, November 1947

The AFL joined in a broad movement to repeal "the Taft-Hartley Slave Law," and created one of the first political action committees for this purpose. The Labor's League for Political Action (LLPE) also worked to increase voter participation, to track state legislation, and to identify like-minded candidates. The Taft-Hartley Act was not repealed, but immediate opposition to it galvanized support to re-elect Truman and elect other Democrats in 1948.

While the IBEW restructured and the labor movement took on its political opposition, the electric power system in the Northwest was still resolving major conflicts between publicly and privately owned utilities.

> From 1940 to 1945, the ratio of public power to private power in the US Northwest fully reversed, with public power making strong gains over the commercial sector.
> Hirt, p. 325

Hirt goes on to describe the geography of publicly owned power.

Rural co-ops dominated the most remote areas; municipal-owned utilities were scattered throughout the region in both large and small cities; and county-wide PUDs dotted the Northwest, competing with pre-established private utilities serving extensive interstate territories.
Hirt, p. 325

After World War II the BPA continued to be the biggest force in supporting the Northwest's preponderance of publicly owned electrical power. By the end of 1945 Congress had given the BPA the authority to market power from six new projects, including McNary Dam (on the Columbia River) and the Hungry Horse Project (on the Flathead River in Montana).

The BPA also backed various PUDs as they initiated condemnation actions against investor-owned utilities (IOUs), including those of various Puget Sound Power and Light (PSP&L) properties. BPA's backing of PUD condemnations came at the same time several major regional IOUs restructured. Some of their corporate changes were in reaction to court decisions relating to the federal Public Utilities Holding Company Act (PUHCA), some came in light of BPA leases that would soon be up for renewal, and some came from estimations of future power demands in the Pacific Northwest.

By supporting the PUDs and a new wave of REAs and utility coops, the BPA effectively determined that publicly owned utilities would remain players in the regional system.

While the regional utility industry retooled after the war, #77 continued to represent construction, operation, and maintenance work on electric generation, transmission, and distribution. The local's membership grew steadily as the volume of electricity used and needed expanded. According to the IBEW's records, George Mulkey, #77's dynamic Business Manager, had moved on to work in San Francisco before the war, either for District 9 or directly for the IO. It was probably his "first assistant" A.E. Martin who secured #77's new contracts as the war ended, one with Seattle Transit in 1944, and another with Grays Harbor PUD in 1945.

As soon as the war was over, new REAs were created across the Pacific Northwest, and they gained more than two million customers between 1945 and 1950, according to the Northwest Power and Conservation Council. Of these, #77 signed formal contracts with Inland Empire REA in 1945 and Big Bend Electric Coop in 1948.

To settle BPA jurisdiction between various IBEW locals, four locals signed what is now known as the Four Local Agreement with "Electrical Contractors of Washington and Oregon" (who would later become Northwest Line Constructors) on February 11, 1946. The purpose of this agreement between Portland #B-125, Tacoma #483, Medford (Oregon) #B-659, and Local #B-77 (signed for by Business Representative Earl F. Wyatt) and the "Contractor" was to:

[E]stablish uniform conditions of employment for electrical Workers cleared by the Union to the electrical Contractor for the purpose of doing line construction, reconstruction, maintenance line work, station and cable work, or other electrical work coming properly under the outside branch of the trade and under the jurisdiction of Local Unions B-77, B-125, B-659 and 483.
Agreement Between Spokane Electrical Contractors Assn., Portland Chapter National Electrical Contractors Assn., Seattle Electrical Contractors Assn., and International Brotherhood of Electrical Workers Local Unions No. B-77, Seattle, Wash., B-125 Portland, Oregon, B-659 Medford, Oregon, 483 Tacoma, Wash.

Nine months later, IBEW District 9 Vice President Oscar Harbak, as Secretary of the Columbia Power Trades Council, met with BPA administrator Paul Raver in Portland. (Harbak had attended the 1946 IBEW convention as a #77 delegate.)

The first meeting brought together management and labor in a formal conclave to discuss relationships bearing on closer relations.
The Electrical Worker, November 1947

No specific IBEW locals are mentioned, nor is the steel tower work on BPA's transmission lines, which became a recurring bone of jurisdictional contention between #77 and the ironworkers.

But even as #77 held its own in the expanding power industry, it could not hold its own as a local. In September 1948, #77 requested that the IO come in to manage the local.

Sketchy information relating to this receivership can be found in three sources: the minutes of #77's Executive Board; the IBEW's national magazine, *The Electrical Worker*; and research relating to specific contracts signed by #77 during 1948 and early 1949.

There is no mention in any of these sources as to what triggered this receivership, and it wasn't until March 1949 that *The Electrical Worker* reported on the situation. This issue included the September 3, 1948 IEC minutes reporting #77 had voted to request the IO "retain supervision of Local 77 until the next regular election in June 1949." A lengthy discussion of procedures followed, under the headline "Local Union 77 of Seattle."

> Our Constitution empowers the International President as follows:
> "To take charge of the affairs of any L.U. [Local Union] when in his judgment such is necessary to protect or advance the interests of its members and the I.B.E.W., but for a period not to exceed six months. If the I.P. [International President] or his representative cannot or has not adjusted the affairs of the L.U. involved at the end of this period, then he shall refer the entire case to the I.E.C. which shall render a decision at its next regular meeting. The I.P. may suspend any local officer or member who offers interference in such cases." Art. IV, Sec. 3, Paragraph (9).
>
> President Tracy has repeatedly stated how distasteful it is for him to feel compelled to exercise such authority. He has exercised it only when there was nothing left to do to protect the interests of a Local Union's members and the Brotherhood.
>
> Because of the unfortunate condition in Local Union 77 of Seattle, Washington, charge was taken of its affairs last April. This was done through Vice President Harbak's office.
>
> VOTE OF THE MEMBERS
>
> Matters were greatly improved and certain corrections and changes were made after the International took charge. And September 3, 1948 the Local Union voted to request that "the International retain supervision of Local 77 until the next regular election in June 1949."
>
> The International President, as our law provides, referred the entire case to the Executive Council. After full consideration the Council decided that International supervision be continued until the next regular election for Local Union officers in June 1949- or until further notice.
>
> "Executive Council Meeting," Minutes of the Quarterly Meeting of the International Executive Council, Beginning December 31, 1948 in Washington, D.C.
>
> *The Electrical Worker,* March 1949

The time lag in reporting on this receivership is particularly interesting because the IBEW had held its biennial convention in September 1948. Local #77's delegation was much diminished (twenty-four delegates representing 5,140 members vs. thirty-three delegates representing 3,543 members in 1946). How much of this reduction in delegates was from restructured IBEW rules regarding representation is unknown.

Most longtime #77 leadership, including Irving Pattee (correspondent to *The Electrical Worker* for some ten years*),* H.S. Silvernale (#77 president from the mid-1940s to the mid-1950s), and Earl F. Wyatt (the business representative who signed the Four Local Agreement, the agreement between the NECA and four regional IBEW linemen's or construction locals, #77, #125, #483, and #659), did not attend the 1948 convention. Only nine #77 delegates attended both conventions, including the invincible George Mulkey (who was once again a #77 delegate although he was working at the time in San Francisco for District 9) and Lloyd C. Smith (who would become #77's business manager when it came out of receivership). Local #77 also sent one woman delegate to the 1948 convention, Clara Schmidt. There is no record of her position with #77.

While in receivership, #77 continued to sign new contracts. Despite previous rancorous dealings with nascent PUDs, in 1948 #77 signed with three of them (Cowlitz County, Douglas County, and Ferry County PUDs), following Mulkey's "softening process" of targeting a single employer until it signed with #77.

Also during 1949, 10 percent of PSP&L's customer base was sold to eight County PUDs: Chelan, Douglas, Grant, Mason, Lewis, Cowlitz, Pacific, and Grays Harbor, according to Chelan County's historical records. The Chelan County PUD also acquired distribution properties in Chelan, Douglas, Grant, and part of Kittitas County, with those outside of Chelan County to be transferred to the Douglas County PUD as soon as its first commissioners were seated. No documentation was found as to how #77 membership at these various employers transferred the terms of their union contracts.

Ellensburg City Light signed with #77 after the local was granted a city council hearing in May 1948. Local #77 presented a "Petition in Support of Labor Agreement, Wages, and Conditions of Employment." The petition compared the specific wages, workweek hours, paid holidays, sick leave, and vacation plans from three

*WWP provides coffee from
a large electric percolator.
Avista Archive.*

*Union-sponsored information booth.
University of Washington Libraries,
Special Collections, UW35555.*

*Governor Langlie signing a bill
as an unidentified representative
of #77 looks on, circa 1941–1945.
Local #77 Archive.*

other cities, four PUDs, and one investor-owned utility already represented by #77. It closed by making clear that signing with IBEW or any union would not create an "unlawful surrender or delegation of [municipal] power."

Local #77 was organizing and signing new contracts with PUDs, municipalities, and federal projects, but it is not clear that other IBEW locals were having as much success organizing utility and line construction work.

In December 1948, IBEW Secretary Scott Milne addressed the national convention of the National Electrical Contractors Association (NECA). Milne's speech made it sound as though linemen had only recently joined the IBEW. Perhaps this was because the majority of NECA's dealings with the IBEW, in most regions of the U.S., related to construction wiremen, not linemen.

> For years, the wiremen have predominated so far as NECA is concerned, and for many years, frankly, they have dominated the IBEW, and justly so because there were many more of them in the union.... [F]or years the inside wiremen have done both the inside work and the line work, and you contractors have been used to dealing with wiremen and wiremen's locals, and the line work being a very infinitesimal part of that jurisdiction, you did not worry too much about it....
> The linemen of your industry are just as interested in the making of our organization as the wiremen are... just as interested in seeing that you, as contractors... work to the end that all line work be controlled, first, by the contractors who will receive the contracts and who are members of NECA, and second, that all work will be performed by members of the Brotherhood who work for you as contractors.
> Milne address to NECA Convention, December 1948, in *The Electrical Worker,* March 1949

Milne, himself a former lineman with #125 (Portland), reshaped the IBEW's pitch on "boomer" work to get NECA line contractors to see it was in *their* interest to work with IBEW. Both parties together could keep out non-NECA, non-IBEW contractors and workers. Part of Milne's interest in securing NECA line construction work for IBEW members had to do with his dark view of the future.

> I think we [NECA and IBEW construction work] are in a decline... and it will continue [to decline].
> ...Are you [NECA contractors] thinking about the time when you will have no work for the 100 men or 200 men you employ, who will be sitting on the benches of our unions?
> Milne address to NECA Convention, December 1948, in *The Electrical Worker,* March 1949

Local #77 came out of receivership in June 1949, ready to move on NECA and other work. Before the election, the IO set salaries for the local's business representatives: $100 per week during a representative's first year of employment, and $110 per week thereafter. Members of the local then elected a slate of candidates including Business Manager Lloyd Smith and President H.S. Silvernale, and a woman, May Armstrong. There is no record of her position with #77.

After #77's officers were installed, new bylaws were adopted that "give state-wide voice in the conduction [*sic*] of Local No. 77's affairs. Now all sections have component local units with an equal vote." Local #77's November report to *The Electrical Worker* went on to describe a dues increase of a "flat 80 cent increase across the board" and the addition of two Executive Board positions (for a total of six). The article described #77's jurisdiction:

> Local No. 77 has three major utilities in its jurisdiction, of which the contract with Washington Power was negotiated for another year in July. Seattle City Light's budget is under consideration by the City Council and includes our requests. Puget Sound Power and Light negotiations will start in November with a contract expiring December 31, 1949. In addition, during the next year contracts with smaller units such as public utility districts, independent telephone companies, and radio stations will have to be negotiated.
> *The Electrical Worker,* November 1949

In fact, #77 signed first contracts with four Central Washington PUDs and one Central Washington REA in 1949: Benton County PUD, Benton REA, Chelan County PUD #1, Franklin County PUD (Pasco), and Grant County PUD #2.

Local #77 also secured two contracts on federal and post-military work that same year. In April 1949, as an affiliate of the newly formed Columbia Basin Trades Council (CBTC), the local signed a multi-craft master contract with the U.S. Bureau of Reclamation (USBR).

Another master agreement was signed in 1949 at the Hanford Works (HEW, now known as the Hanford Site). General Electric Company signed as "operator for the U.S. Atomic Energy Commission, of the plants, properties and facilities owned by US Atomic Energy Commission [AEC]." (The AEC had taken over the administration of Hanford in 1947.)

On the union side, Hanford Atomic Metal Trades Council (HAMTC) was the exclusive bargaining agent for its affiliate union locals and the "international unions who are themselves affiliates," each of which represented a distinct group of craft employees. (Formed in 1949, HAMTC affiliates included #77, IBEW #112 Sound and Communications, the Teamsters, and unions representing insulators, ironworkers, machinists, operating engineers, painters, plumbers and pipefitters, steelworkers, sheet metal workers, boilermakers, and instrument technicians).

1949 was also the year that the Washington State Power Commission (WSPC) was created.

> For decades, no [Washington State Legislative] session had been complete without its own version of the public-private power debate. This time [during the 1948 session] it centered upon a three-part proposal pushed by the public power community. It included the creation of a state power commission, for P.U.D. tax equalization, and most significantly to allow P.U.D's to jointly acquire the properties of private utilities, an action which had recently been found improper by the Supreme Court of the state. The proposal was not without substantial controversy, however, it did pass and was signed by Governor Langlie.
> Don Brazier, *History of the Washington Legislature 1854–1963*, p. 144

Note that PUDs were explicitly allowed to "jointly acquire the properties of private utilities." This language was designed to allow six PUDs to pursue a very specific business transaction:

> to jointly purchase the remaining Puget Power properties under arrangements being negotiated by Guy Meyers.... The joint action law was passed but it was limited.... There would not be any joint condemnation of private [investor-owned] power properties.
> Ken Billington, *People, Politics, and Public Power*, p. 38

Political and economic allegiances profoundly shape the various accounts and omissions of events within Washington State's utility industry from the late 1940s into the 1960s. Historians disagree about whether the fight for public ownership of the electrical industry had been resolved by the early 1950s or whether the fight was entering its critical phase. Hirt sides with BPA's historian, Gene Tollefson, that "the Northwest Power Wars" were over by 1952.

Billington, coming from the PUD perspective, devotes two chapters and more than 150 pages to documenting the legal and political wrangling over public and private ownership of hydropower infrastructure in Washington State from 1951 to 1963.

Because #77 sought to represent utility workers across the state, it took an approach similar to that of the IBEW when the International faced the Giant Power Movement. Local #77 cared less about the type of utility ownership than it did that all utility workers were union. Using its tried and true industrial model, #77 picked up maintenance and operation work as construction work sometimes boomed and sometimes busted. Local #77 continued to organize all the workers of an employer whenever possible, and pegged everyone's wages to those of the lineman.

Coworkers of Irving Pattee, a longtime #77 correspondent to The Electrical Worker, *somewhere near Marblemount, Washington. Ballock standing, second from left. Kneeling from left: Wilder, Todhunter, unknown, Hagen. Local #77 Archive.*

"The Supreme Court," a SCL crew in the Ross Dam powerhouse during cavitation (the process of removing bad metal, welding new metal to replace it, then grinding the turbine down to correct contours.) L-R seated: Pete Gasier, Don Haberkamp (on Pete's knee), Charles Smith, Gary Moore. L-R standing: Mack Moore, Daryl Ginther. Gary Moore Archive.

Modern Leaders of Local #77

FROM 1950 until today, #77 has remained a player as the utility industry, labor law, and corporate law underwent dramatic changes. Local #77's ability to survive and thrive has had much to do with the personalities and priorities of each elected business manager. Because of their vital role, the second half of this history will document the events and actions of each business manager's term.

Opposite page: A variety of PSP&L meters.
1) Westinghouse Type B, single phase.
2) Westinghouse Type OA, 2 wire, single phase.
3) Westinghouse Type OB, 3 wire, single phase.
4) Westinghouse Type OC, 2 wire, single phase.
5) Westinghouse Type CA, 2 wire, single phase.
6) Westinghouse Type OC, 2 wire, single phase.
7) Thompson DC, Type M.
8) Thompson DC, Type M.
9) Thompson DC, Type M.
10) Shallenberger ampere-hour meter.
11) Stanley, single phase.
12) Thompson polyphase, 3 wire.
13) Westinghouse Round Type, polyphase.
14) Westinghouse Type B, single phase.
15) Westinghouse Type C, 3 wire, single phase.
16) Westinghouse Type polyphase.
17) Westinghouse Type, single phase.
18) Westinghouse Type, single phase.
19) Stanley, single phase.
20) Westinghouse Type C, polyphase.
21) Ft. Wayne, Type K, single phase.
University of Washington Libraries, Special Collections, UW35564.

Above: The PSP&L board of directors signs off on SCL acquisition of PSP&L holdings. University of Washington Libraries, Special Collections, UW35565.

Below: Lloyd Smith, October 1954. The Electrical Worker, October 1954. IBEW IO Archive.

Below right: Separate power lines serving the same area, labeled City Light (SCL) and Private Company (PSP&L). University of Washington Libraries, Special Collections, UW35599.

C. Lloyd Smith
Business Manager
1949–1955

A slate of candidates, including Business Manager C. Lloyd Smith, was elected when #77 came out of receivership in mid-1949. Smith continued as #77's business manager until 1955, leading #77 during immense changes in the Pacific Northwest power industry.

As the U.S. shifted into a postwar economy, so did state and federal law concerning electrical generation, transmission, distribution, and the structure of utility corporations. In Washington State the battle between publicly owned and investor-owned utilities (IOUs) picked up where it had left off before World War II.

> If a one-word description of the private/public power war during the 1951–1956 era were needed, it would be "leveling" or perhaps "equalizing."
> Ken Billington, *People, Politics, and Public Power*, pp. 128–129

The details of this "power war" are beyond the scope of this book but a brief summary of the corporate and legislative changes of the first decade after World War II is worth noting.

In 1942 the U.S. Securities and Exchange Commission (SEC) had directed utility holding companies to divest themselves of what Billington calls "their captive private power companies." In response, the American Power and Light Company (AP&L), owner of both Washington Water Power (WWP) and Puget Sound Power & Light Company (PSP&L), tried to merge the two companies in 1947.

When the SEC denied this proposed merger, AP&L sold all of PSP&L's common stock to "a private underwriting group who could rightfully be called speculators," according to Billington. This group held PSP&L's stock "just long enough" to avoid paying capital gains taxes, then sold to new owners about a year later. Its stock value jumped more than one hundred percent in this one year.

Between 1948 and 1952 there were more deals and proposed deals for purchase of PSP&L holdings than can easily be listed. Major offers included the City of Tacoma's proposed condemnation of PSP&L generation properties in 1948. Various PUDs had been attempting condemnations of PSP&L holdings since before World War II and continued to do so. Lewis and Chelan County PUDs bought PSP&L distribution facilities in 1948 as did the Snohomish PUD in 1949. Shortly after this Gus Myers, an ex-Wall Streeter who worked as the fiscal agent of various PUDs, made a buyout offer to PSP&L on behalf of a group of PUDs and Seattle City Light (SCL).

In 1950 SCL dropped out of this joint offer and made a separate offer on the PSP&L properties within Seattle. (The City of Seattle had passed a resolution in 1943 to take over PSP&L's Seattle holdings before the city's fifty-year lease with PSP&L expired in 1952.) Seattle's stand-alone bid was accepted by PSP&L and went to Seattle voters as Proposition C in November 1950, passing by only a few hundred votes. PSP&L and Seattle signed the deal on March 5, 1951. Seattle paid approximately $27 million for 40 percent of PSP&L's revenue base, according to *A Century of Service,* PSP&L's corporate history. PSP&L's steam system spun off into a separate corporation, Seattle Steam Company.

> With the final sale of property to City Light, one company broke away, Seattle Steam [Company]. A group of local businessmen and steam users stepped in and bought the steam production facilities at Post Avenue and Western Avenue.
> Amanda Roberson, "An Investigation of Post Avenue Steam Plants," 2003, p. 5.

Not much more than a month after PSP&L and SCL's deal, Chelan County PUD entered into what *A Century of Service* calls a "marriage of convenience" with PSP&L, wherein PSP&L agreed to purchase the majority of power generated by the proposed Rock Island Dam for the next fifty years if Chelan PUD financed its federally mandated construction. This was in response to rumors that President Harry S. Truman was ready to terminate the Rock Island Dam permit.

Meantime, AP&L needed to divest itself of WWP. Guy Myers made an offer on behalf of the Chelan, Stevens, and Pend Oreille PUDs. This offer proposed that these PUDs would get all the common stock of WWP while WWP property in Idaho would go to an Idaho nonprofit.

> AP&L decided that it wanted to cut its losses and divest itself of WWP, even if the company had to be sold to a public utility district (PUD). WWP's board disagreed with this decision, and even garnered the support of some Spokane businessmen who were then able to get a restraining order to temporarily prevent the sale.
> The case [whether WWP could sell to the PUDs] was tried in March 1951. The decision handed down made it clear that the sale of WWP could not be legal because the company had interests in several states, and PUDs were not able to own property in states other than the one in which they operated.
> Lehman Brothers Collection website, "The Washington Water Power Company," 2003

Billington portrays this offer differently. He describes political machinations during the 1951 Washington State Legislature, including a proposed "Spokane Power Bill" that would have put any PUD's purchase of WWP to a public vote in Spokane. This bill seems to have had "the Spokane businessmen" described above as backers. At any rate, AP&L did divest itself of WWP.

> In 1952 a new board of directors was formed consisting of members selected by both WWP and AP&L. Shares of WWP stock were distributed to AP&L stockholders, and for the first time since William A. White rescued the company from bankruptcy in 1895, WWP was on its own.
> Lehman Bros Collection, "The Washington Water Power Company," 2003

Then came what can only be described as backroom dealings. In September 1952, six PUDS (Chelan, Kitsap, Jefferson, Skagit, Snohomish, and Thurston) made an offer on all of PSP&L's properties except those in Whatcom County. (Whatcom County PUD, originally a partner in this offer, had withdrawn.) A majority of PSP&L's stockholders voted to accept.

The PUDs had four months to raise funding. During this time WWP's CEO, Kinsey Robinson, presented a shifting array of offers to PSP&L. First he proposed that WWP and PSP&L merge, and secured an "understanding" between the two corporate boards to do so. At the same time Kinsey was also working to block the PUD sale.

The February 1953 closing date on the PUD sale was extended in a maelstrom of legal actions between individual and collective PUDs, PSP&L, and WWP that stretched into the fall of 1953. Meanwhile public opinion continued to support the PUD offer, but voters did not want Washington to end up with a single, huge IOU (investor-owned utility). It also became clear during Public Service Commission hearings that PSP&L could probably stand on its own financially without selling to the PUDs or merging with WWP. "After multiple courtroom battles the Puget Board of Directors in November 12, 1953 decided not to merge or sell Puget properties," summarizes the Chelan County PUD history.

While PSP&L and WWP were bargaining, there was a shift in federal policy regarding federal financing and control of public power.

With the election of Eisenhower, the so-called partnership power policy started to evolve.... this policy was to reduce federal participation in power development, and since the federal power involvement in the Pacific Northwest was large, the policy did not bode well for public power people.
Billington, *People, Politics, and Public Power*, p. 73

Tollefson describes Eisenhower's "partnership plan" for the development of hydropower between public and private corporations as a way of getting the federal government out of paying for future steam generation. (Increased steam and coal generation was proposed to cover "brownouts" when Northwest river levels dropped, as they had in 1952.)

BPA Administrator Paul Raver, who had been a public-power loyalist since his 1939 appointment, was directed to offer long-term power contracts to IOUs, and to put the IOUs ahead of any new industrial customers when allocating hydropower. Raver had offered long-term contracts to the IOUs in years past. These had been rejected by the IOUs because of specific rate regulations and "purchase" provisions, according to Tollefson.

In 1953 Dr. William Pearl replaced Raver as BPA administrator. Almost immediately Pearl fired 600 of the 2,700 BPA employees. Most of these were construction linemen (probably members of IBEW #125, out of Portland). Tollefson quotes Price:

Bonneville was doing an awful lot of its own construction work.... there was an opportunity here to reorganize and redirect the Bonneville effort... utilizing private forces.
Tollefson, p. 291

In other words, BPA privatized much of its line construction.

At the state level, the 1953 legislative session featured what Billington calls the "million dollar babies of private power," two anti-PUD bills. House Bill 77 and Senate Bill 54 proposed prohibiting PUD condemnation of privately owned property through procedural actions that would be impossible for a PUD. (Chelan County was moving to condemn WWP's Chelan Falls generating plant during this period.) In addition, other draft bills proposed to block future PUD purchases of PSP&L assets. None of these passed.

A state law allowing publicly owned utilities to form joint operating agencies did come out of the 1953 session. This would be used to create the Washington Public Power Supply System, or WPPSS.

There is little documentation that speaks directly to #77's political involvement during the late 1940s and early 1950s. It would seem that the local focused on organizing and holding onto contracts no matter what entity became the employer.

A case in point is the purchase by Seattle City Light (SCL) of PSP&L's property within Seattle's city limits. Before the purchase, #77 represented workers at both SCL and PSP&L, under separate contracts. These #77 members were "merged" and contract terms including job classifications, rates of pay, benefits, specific conditions of work, and seniority had to be combined as well. Lou Walter, now a business representative for #77, remembers "Puget guys" at SCL, who had separate dues records.

Local #77 representation at the Seattle Steam Company also evolved from SCL's buyout of PSP&L properties. Charlie Munson relates what he heard from old-timers at Seattle Steam. (Munson is now Seattle Steam's manager of distribution and customer service.) When Seattle Steam became autonomous it inherited contracts with multiple unions in its co-generating shop. Consequently, some contract or another was almost always in negotiation. Over time, about when Richard McKay became Seattle Steam's manager, the company decided it wanted to have one union represent all its workers. Munson heard that #77 was chosen because it was the best fit. For some period after Seattle Steam signed with #77, its negotiations generally "piggybacked" on those of SCL.

During Smith's time as business manager, #77 signed approximately fifty new contracts. (This figure is based on contracts on file at the #77 offices and in the University of Washington Special Collections.) Some of these first contracts formalized existing #77 representation (including telephone workers and Private Branch Exchange, or PBX workers).

Other contracts reached out into radio and the first TV broadcasting. In the late 1940s and early 1950s, many of #77's new contracts were with PUDs as these utilities began operation. For example, Snohomish PUD purchased PSP&L's distribution system with the county in 1949, after more than ten years of wrangling with PSP&L. Months later #77 was at the table with Snohomish PUD.

Clockwise from left: Local #77 picnic, 1953. Local #77 Archive.

Stranded bear cub being rescued by Don Frost, 1950s. Frost was a lineman in Pullman for WWP. In the early 1960s, he became a #77 business rep in Palouse, Washington. Avista Archive.

Comparison of IBEW member wages, 1947–1954. The Electrical Worker, October 1954. IBEW IO Archive.

Re-insulating the highline to Julietta [sic]: George Bockman (lowest on pole), Don Frost (upper pole), 1952. Avista Archive.

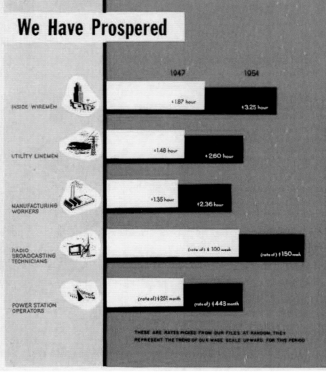

Negotiations with numerous P.U.D.'s are about to start and more particularly with the new Snohomish P.U.D. which was recently formed by purchase from Puget Sound Power and Light.
The Electrical Worker, February 1950, #77 report

In February 1953 #77, in partnership with NECA and possibly some IBEW locals in Oregon, began a formal apprenticeship program. From then on, an apprenticeship class has graduated from this program every year.

[The] first two apprenticeship classes were held in March; [apprentices] required to attend class one Saturday a month in Seattle with approximately twelve hours a month of homework. Tuition fee is $25.00 which includes all books, etc.
Report to H. Conover from business representative D.J. Kleinman March 30, 1954

By summer of 1955, #77 had moved to a new office, built for the local. Located on Melrose Avenue, some eight blocks from the Labor Temple on First Avenue, it seemed far away from "labor row," the section of Belltown close to the Labor Temple where many unions had their offices.

In 1955, after an acrimonious campaign, Lloyd Smith did not win re-election. Details of Smith's fight to remain business manager are detailed in the next chapter.

The 9th District Delegation to the 25th IBEW AF of L, August 30 to September 4, 1954, at the International Amphitheatre, Chicago, Illinois. Note banner in the background, "Where Electricity Goes, There Goes the IBEW." Local #77 sign is in the second to the back row, center. The Electrical Worker, *October 1954. IBEW IO Archive.*

Above: Local #77 dam operators' banquet, Grand Coulee Dam April 27, 1956. Local #77 Archive.

Left: IBEW/AFL button. The AFL and CIO merged in 1955. Rick Luiten Collection.

Below: Private and public ownership of generating facilities was in flux during Henry Conover's term as #77 business manager. On November 29, 1955, WWP issued a press release including this photograph. Stevens County voters had approved the sale of Stevens County PUD properties to WWP. The PUD had moved to condemn WWP properties in Stevens County and WWP conducted a major political campaign to instead shift the PUD property to private ownership. This check was for $2,905, the balance due from WWP for all of the Stevens County PUD facilities and properties. Northwest Museum of Arts & Culture/Eastern Washington State Historical Society, L2007-1.17.6.5.

Henry M. Conover
Business Manager
1955–1958

HENRY Conover came out of the Spokane office of #77, where he had been a business rep from 1949 until the summer of 1952. IBEW records indicate that he was a "special organizer" for the IO from August 1952 to August 1953, but make no mention of what this work involved. Conover was back in Spokane working for #77 in 1953. When he declared he was running for business manager on March 18, 1955, then-manager Lloyd Smith fired him by phone, and documented the firing in a letter to Conover.

Two weeks later Conover wrote his own letter to #77's executive board, requesting two weeks severance pay:

> I was terminated as a representative from the business staff of Local 77 because of my intention to be a candidate for the office of business manager.
> Conover letter to J.F. Flynn, #77 recording secretary, April 1, 1955

Flynn asked for clarification from the IBEW. IBEW District 9 Vice President Oscar Harbak (a former #77 member) responded that severance pay was officially a matter of local policy, and that a precedent had been set by #77, so the local "would not be out of line in approving severance pay in Henry Conover's case."

Conover's election was not a friendly shift in power. Once elected, Conover immediately fired five #77 business representatives because they were "political appointees;" Dave Kleinman, Jake Gilbert, Aubrey Hart, Gordon Smith, and Gene Nelson.

Conover gave the staff he kept and hired his full support, however. Bill Shaffer, a longtime #77 business rep in Central Washington who worked with Conover, said, "Nobody compared with Hank Conover. He knew how to do it." Shaffer explained, "Hank gave me free rein while he worked like a machine. Unstoppable." Others who knew Conover described him as more than able to enforce his own decisions and opinions.

Records do not document any new employer contracts during Conover's term. Local #77 seems to have had its hands full retooling existing contracts with both recently restructured employers WWP and PSP&L, and SCL as it adjusted to its recently acquired infrastructure, equipment, and personnel. In addition, #77 signed contracts with the new PUDs and REAs as they became more fully operational. Former business rep Bill Shaffer recalled, "In the 1950s, the [employer] transitions weren't bad, things went OK."

While #77 was catching up on its contracts, plans for new hydropower development were proposed from all quarters to meet a regional need for more electricity. Three organizations were created to make these proposals fit the federal "partnership" policy on new power generation.

In 1953 the Puget Sound Utilities Council began to meet, although it did not incorporate. Its members were a mix of publicly and privately owned utilities: PSP&L, SCL, Tacoma City Light, Snohomish County PUD, and Chelan County PUD.

In 1954 a private consortium, the Pacific Northwest Power Company (PNPC), incorporated so its members could jointly build and operate hydroelectric and other generating plants. At the time of incorporation it was made up of five private corporations: WWP, Portland General Electric, Montana Power Company, PSP&L, and Mountain States Power. Mountain States Power merged with PSP&L soon after the PNPC was created, almost doubling PSP&L's service area. Local #77 took on many former Mountain States Power employees under its PSP&L contract.

While WWP was pursuing plans for new generation within the PNPC in 1955, WWP's 48-megawatt Chelan hydro project was "condemned, a victim of the public power movement," according to a historical account on Avista's website (WWP became Avista in 1998). The Chelan County PUD describes this transfer:

> June 21, 1955 Acquired distribution system in Lake Chelan area and Chelan Falls Hydro Plant (Lake Chelan) from WWP; District entered into 40 year power sales contract with WWP.
> Chelan County PUD, "History of Events Affecting Chelan County PUD"

Shaffer recalled #77 was not involved in discussions between Chelan PUD and WWP as the Rocky Reach and Chelan Falls plants "went to" Chelan PUD. He can remember workers who "swung over" from WWP to the Chelan County and other PUDs, and that the transition "wasn't a big deal" for #77.

A third utility organization was created in 1957, the Washington Public Power Supply System (WPPSS). Seventeen public utility districts organized "in order to combine their resources to build generation." (Public Power Council, "Public Power History")

While Washington State utilities formed these various alliances to build new generating plants, #77 "followed the electricity" to hold onto contracts with publicly and privately owned utilities. It continued to use an industrial model, organizing all the workers and all the job classifications of any one employer.

Nationally, the IBEW was using a similar industrial model. According to Palladino, "Seventy-five percent of (national) IBEW membership in 1954 was in industrial locals." The IBEW success included utility workers across the U.S.

> [The IBEW] organized seventy-five percent of the nation's utility workers (most of them private, not public, utilities)... as well as making inroads in the telephone and manufacturing industries, with a membership of 625,000 in more than 1,675 locals, making it one of the AFL's largest affiliates.
> Palladino, p. 217

Consequently, the 1954 Chicago IBEW Convention was the largest labor union convention ever held in the world. J. Scott Milne, who had topped out as a lineman (made journeyman) in Portland's #125, was elected president. The 1954 convention focused on restructuring, creating separate industrial departments (including manufacturing, utility, telephone, and construction), in part to address rumblings and discord over jurisdictions between IBEW locals.

In reality, the IO had bigger jurisdictional problems to deal with. The AFL and the CIO had signed a non-raiding agreement in 1953, and merged in 1955 to form the AFL-CIO. No record was found of #77's positions on either the 1953 non-raiding agreement, nor any #77 participation either at the 1954 convention, as the IBEW got ready for the IBEW AFL-CIO merger. The November 1955 *Electrical Worker* paraphrased AFL President George Meany, who emphasized that "there would be cooperation and understanding among unions where jurisdictions paralleled."

The IBEW would have a front-row seat for seeing how this "parallel jurisdiction" policy played out, as the International Union of Electrical, Radio and Machine Workers (IUE-CIO, a descendant of UE, the United Electrical Radio and Machine Workers of America) was a major player in the CIO and in the merger.

Conover left #77 as the AFL-CIO merger became reality. He was appointed director of IBEW utilities operations (now the utilities division) at the end of 1957.

Above: Henry Conover became business manager months after #77 had moved to a new office built for the local. Located on Seattle's Melrose Avenue, some eight blocks from the Labor Temple on First Avenue, it seemed far away from "labor row," the section of Belltown where many unions had their offices. Local #77 Archive.

Right: Neon sign on #77's Melrose union hall, probably made and installed by #46 members. Local #77 Archive.

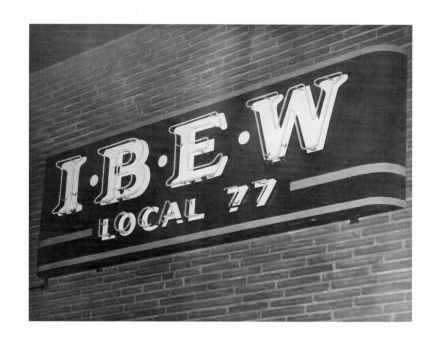

The McClellan Committee hearings of May 1957 included Robert F. Kennedy, committee counsel, and Senator John F. Kennedy of Massachusetts. In the wake of the hearing, Congress enacted the Labor-Management Reporting and Disclosure Act (Landrum-Griffin Act) in 1959. University of Denver, Sturm College of Law website.

Above: A KING-TV cameraman (a #77 member) covering an unknown event at the new #77 hall on Melrose Avenue in Seattle. When Kenny came on, much of the local's growth came from organizing broadcasting companies. Local #77's report to the June 1963 issue of The Electrical Worker *reported, "Local 77 now has the following stations on the air und under contract: KING-TV Seattle, KTNT-TV Tacoma, KHQ-TV and KXLY-TV, Spokane. The first TV Station contract, that of KING-TV, has been printed and is available. The new KING contract calls for the following scale: engineers, $2.84, crew chiefs $3.10, and supervisor's $3.30. This is tops for the present since with an established market and time charges of $640 an hour, KING-TV roles the roost for revenue in the Northwest....Clem Seeber, Local 77's radio representative and W. A. Smith of the International have over 400 members and more than a contract a week to keep them out of mischief." In 1968, #77 lost its broadcast units, which were transferred to IBEW locals #73, #112, #497, #191, #46, and #76. Local #77 Archive.*

Local #77 joined IBEW locals #483, #48, #49, #112, #73, #191, #970, #125, #497, #46, and #76 in hosting the 1970 IBEW International Convention is Seattle. Forty-five years earlier, in 1925, Seattle had hosted the 18th IBEW International Convention. Local #77 Archive.

IBEW 29TH CONVENTION PROGRAM

Seattle, Washington
Sept. 28 - Oct. 3, 1970

Arthur B. Kenny
Business Manager
1958–1971

WHEN Arthur B. Kenny was appointed business manager of #77 at the beginning of 1958, #77 was growing steadily, with a membership of almost 6,500. It appears much of this growth came from organizing broadcasting companies.

Consumer demand for energy continued to surge nationally and regionally. The first major natural gas pipeline came to the Pacific Northwest in 1956, and within three years natural gas had become an exponentially growing competitor to hydropower. WWP took on the region's major natural gas supplier, Washington Natural Gas Company, by diversifying, and in 1958 purchased its first natural gas company, Spokane Natural Gas.

At the national level, the IBEW was growing rapidly. Gordon Freeman was appointed IBEW President when Scott Milne died suddenly in 1955. Freeman, a wireman who had helped organize the Tennessee Valley Authority (TVA) using an industrial model, wanted the membership to focus on the growing economy, and related growth within their craft. IBEW members "had averaged wage increases of at least twenty percent since 1954." (Palladino, p. 222) Building construction was booming and so was manufacturing. Electrical consumption was exploding, and this would drive the expansion of utility infrastructure.

Between the 1954 and 1958 conventions, the IBEW's membership increased twenty percent, according to Palladino. But this growth came largely from the boom in building construction rather than an increase in linemen's or utility work.

IBEW's 1958 convention was contentious, and focused on the IO's inter-local jurisdictional policies. Palladino summarizes the IO's position going into the 1958 convention:

> Ever since the 1930s, the Brotherhood had assumed a rather common-sense approach to the issue: the local that organized a particular job and successfully defended its claim to the work was the local that had jurisdiction.
> Palladino p. 221

In 1958 many IBEW locals were successfully organizing outside of their traditional jurisdictions. As the U.S. economy grew, inside locals (wiremen) were going after manufacturing work, and #77 wasn't the only utility and construction local that was bringing in radio and TV workers.

Art Kenny and the other #77 delegates came back to Washington State from the 1958 convention and joined other state labor leaders in taking on a more immediate challenge. Initiative 202, "Restricting Labor Agreements," was on the ballot. Voters rejected it, and Washington did not become the nineteenth state to pass right-to-work legislation.

The following year, 1959, saw passage of a federal law regarding union activities, the Labor Management Reporting and Disclosures Act (better known as the Landrum-Griffin Act). Many labor leaders supported the bill's anti-corruption and anti-racketeering provisions, but the Landrum-Griffin Act further restricts strikes, boycotts, and picketing, and also requires unions to file detailed annual reports.

With a surfeit of work, the IBEW became both complacent about organizing smaller construction jobs and protective of the best-paying construction work.

> Work was so plentiful in some locations that union electricians bypassed residential markets completely, concentrating on large, more lucrative commercial projects. At the same time, these locals continued their restrictive membership policies, issuing work permits on a temporary basis and tightly controlling entrance to the trade, actions that played right into the hands of right-to-work proponents and ultimately strengthened the nonunion sector of the construction industry.
> Palladino, p. 215

Going into the 1960s, the IBEW took a similar approach toward expanding work in nuclear power, federal work related to the Cold War such as manufacturing military vehicles and munitions, and defense installations like the Distant Early Warning Line (DEW, or DEW line), an extensive series of radar stations in Alaska.

Even as federal work continued to boom, construction slowed and factories lost jobs to automation. The IBEW's complacency allowed non-union contractors to pick up work IBEW locals did not consider worth organizing.

The way actual tasks were performed by linemen was also changing. Virtually every lineman who came up during the 1960s identifies the bucket (hydraulic lift) truck as the single biggest change they experienced in their careers. For the first time, a lineman had an alternative to climbing poles. Hydraulics also changed how poles were set. Rather than a crew piking a pole or using truck-mounted structures and/or winches, a hydraulic lift positioned the pole.

It took awhile for employers to recognize the value of these improvements, and it took linemen longer. "A lot of guys wouldn't use a bucket at first. They thought it was the beginning of the end," recounts Lou Walter, a #77 business representative. To an outsider, this attitude might seem Luddite. To someone who understands a lineman's craft, it is clear how much linework is well-served by being on the pole, no matter what technologies may be available. (Even today, industry statistics cite 40 to 60 percent of current linework done with hooks, climbing, rather than from buckets.)

And linemen weren't the only members of #77 whose work was radically altered by changing technologies. Meter readers were asked to wear machines that could directly read meters and then print out a receipt. Several workers who wore these machines can remember how they couldn't see their feet over the machine, and that the ink on the receipts would dissolve in the rain.

Dispatchers, who had traditionally taken calls from customers and then directed linemen to specific power outages, began the shift from large paper maps to partially computerized representations of the power grid. Customer service representatives (CSRs) began to work with automated call systems, and a CSR's authority to resolve various customer issues began to vary according to what a specific employer decided was most cost-efficient.

As the work itself changed, so did #77's jurisdiction. In 1960 Local #882 (Shelton) was amalgamated into Local #77. The PUDs had become a more solidified group of employers and began trying to negotiate with #77 as a bloc.

> PUD negotiations are complicated this year due to the Public Utility districts insisting on a statewide contract. It doesn't leave much to report because the situation changes from day to day.
> *The Electrical Worker,* February 1960, Local #77 report

The PUDs didn't secure a unified contract in 1960, but by 1962 the PUDs of Central Washington were individually settling for one wage increase (fourteen cents per hour) and their five-year contracts came up concurrently.

By 1963, #77 realized it might be beneficial to try regionally packaging some negotiation with PUDs.

A few years ago Local 77 was faced with a demand from the Washington State Public Utility District Association that negotiations be handled on a state-wide basis. Our membership is completely opposed to such type of contract but it was finally agreed that four of the Western Washington Public Utility Districts, Snohomish, Cowlitz, Grays Harbor, and Lewis, would try it on wages alone with conditions being negotiated separately first.
The Electrical Worker, April 1963, Local #77 report

A year later, the PUDs and #77 signed an agreement that set terms (reimbursement of wages and some work rules) for workers at various PUDs if they responded to a call out from another PUD during a disaster.

The effects of the Columbus Day Storm of 1962 are still being felt. Recently the Washington Public Utility Districts got together and signed a Disaster Agreement which is a direct result of problems which arose in the tremendous job of restoring service to normal after that storm. Under this agreement if one of the PUD's need help from another the employee's wages and conditions will be paid by his own employer and billed to the other PUD. It will be on a voluntary basis and no employee need work for any other PUD unless he wants to.
The Electrical Worker, May–June 1964, Local #77 report

The construction linemen in #77 were not as interested in further standardizing their work nor in changes to their apprenticeship program. Research for this book revealed a number of second- and third-hand accounts or references to a wildcat strike in the early 1960s. No specific documentation was found. Accounts reference some combination of this wildcat strike with a possible attempt by construction linemen to form their own independent IBEW local. Some versions align the wildcat strike and/or the attempt to secede from #77 with a backlash to stepped-up apprenticeship requirements for linemen.

Bill Stone, training director for the Northwest Line Joint Apprenticeship and Training Committee (JATC) and avid historian of the apprenticeship programs, has also heard stories about these events. Stone relates that at the time, more NECA contractors were interested in picking up distribution and transmission work, in addition to line construction work. Many "boomer" linemen had traditionally worked primarily in constructing lines, and did not necessarily have the training or experience to work on electrified, or "hot", lines.

Having apprentices who arrived onsite ready to work hot was a threat to some longtime journeymen. That apprentices were being trained, not by journeymen on the job, but in school somewhere, was not in keeping with the linemen's tradition of hands-on, in-the-field training. In addition, the apprenticeship program was adding and beginning to enforce requirements that apprentices work for several employers rather than with a single employer, and perhaps a single crew. This meant apprentices appeared at and were re-assigned to worksites, rather than coming into their trade under a single group of journeymen linemen.

Part of the difficulty in documenting these events is that whoever may have led or participated in an attempt to form an independent local seems to have been "blackballed," their IBEW membership rescinded. Some linemen who apprenticed in the late 1960s and early 1970s recall certain very respected journeymen linemen who couldn't attend union meetings, who may have been part of this blackballed group. (Certified non-union journeymen can be dispatched by a union, and work alongside union linemen, without being actual members of IBEW.)

During this same period, the early 1960s, the IO had begun to sign major project agreements, with no-strike clauses, superseding the ability of any local, including #77, to negotiate directly with employers.

Right-to-work proponents rallied in Washington State in 1963. Local #77 took action in that legislative session, focusing on five proposed bills, and scoring a "batting average of 3 out of 5, which isn't bad in this league," as #77 reported to *The Electrical Worker*.

Here in the State of Washington, the so-called "Right-To-Work" proponents have twice managed to place this type of legislation on the ballot and both times it was soundly defeated....

Many bills directly affecting our members were introduced, some of them favorable, some of them not. Business Manager Art Kenny worked like a trooper following their progress and keeping the membership informed. The members did a big part by writing, phoning and telegraphing their legislators to solicit their support. All of this paid off.

Raising and repair of damaged 26 k.v. submarine cable. Anacortes Washington, San Juan Islands, July–August, 1959.
Above (L-R): Tommy Roselli (SCL), Fred Campbell, and "Bud" Dillard inspecting armor damage. Local #77 Archive, gift of Brother Tommy Roselli.
Below: Taking a break. Local #77 Archive, gift of Brother Tommy Roselli.

There were five main bills we were concerned with which directly affected our membership. Two of them we opposed and three of them we favored. The two we opposed both failed. One was Senate Bill 454 which was a Bill to amend our 1918 State Electrical Safety Law and would have probably resulted in all our safety rules becoming directive law rather than legislative law and therefore subject to change at any time and lacking in the enforcement power that legislative law has.
The Electrical Worker, May–June 1963, Local #77 report

Local #77 also secured a major change in collective bargaining rights for PUD employees.

> Senate Bill 233 was passed and it is short enough and of such importance that it is quoted here in full:
> Be it enacted by the Legislature of the State of Washington:
> New Section: Section 1. There is added to Chapter 54.04 RCW a new section to read as follows:
>> Employees of public utility districts are hereby authorized and entitled to enter into collective bargaining relations with their employers with all the rights and privileges incident thereto as are accorded to similar employees in private industry.
> New Section: Section 2. There is added to Chapter 54.05 RCW a new section to read as follows:
>> Any public utility district may enter into collective bargaining relations with its employees in the same manner that a private employer might do and may agree to be bound by the result of such collective bargaining.
> Senate Bill 184 was similar to Senate Bill 283 in that it extended these same rights to municipalities. The opposition was such that it lost....
> The passage of Senate Bill 233 brings these PUD [employees] out of that "gray world" of uncertain status that so many Government employees come under. It is also in line with President Kennedy's Executive Order No. 10988 which was signed by the President on January 17, 1962 and went into effect July 1, 1962 and gave these same rights of collective bargaining to Federal [employees].
> *The Electrical Worker,* May-June 1963, #77 report

New federal laws brought other changes to the workplace. In 1963 Congress passed the Equal Pay Act (EPA), amending the Fair Labor Standards Act to prohibit wage differentials based on gender for workers covered by the Fair Labor Standards Act. A year later the Civil Rights Act was passed, landmark legislation that outlaws blatant employment discrimination, prohibits unequal application of voter registration requirements, bans discrimination in public accommodations, and encourages school desegregation, although powers of enforcement were initially weak.

The times were a-changing. In August 1964, the United States stepped up its troop involvement in Vietnam by passing the Gulf of Tonkin Resolution without formally declaring a war. At home, the Labor Law Study Group (LLSG) was formed in 1965. Its goal, writes Sharon Beder, was to oppose unions and weaken labor laws, with the help of the public relations firm Hill & Knowlton.

> The LLSG was made up of twelve "thought leaders," men who were top corporate labour relations executives from the largest corporations and belonged to "all the trade associations in every nook and cranny in the country." They were known as "the Twelve Apostles." By 1968 the campaign was being described as the "broadest united front of large and small businesses in history."
> The "Business Managed Democracy" website, Sharon Beder, "History of the Business Roundtable 3"

Over the next four years, the LLSG would merge with other corporate interest groups to form the Business Round Table. One of the major concerns of the Round Table was the increasing cost of construction, which the group attributed in large part to union wages.

Local #77 met opposition at the negotiating table with strikes. In 1965, #77 members walked out at Hanford. Bill Shaffer, #77's Central Washington business rep at the time, remembered that the local put the striking members out on tree crews and whatever other work was available. "We had them stacking steel, working in the building trades, digging trenches, painting the towers. They did it all. We made sure our members survived." Shaffer punctuated this memory with the oft-repeated #77 refrain, "We were never a country-club union." Local #77 looked out for its members, and its members were willing to take whatever work there was.

Partially because of #77's willingness to represent a variety or work, IBEW #1958 (Richland) amalgamated into #77 in 1964, bringing along its 215 members, all employed at Hanford. Gene Langdell, who became a business rep for #77, was a member of #1958 as it negotiated the amalgamation. Langdell had come up in #1958 as it first organized with the Instrument Technicians Guild as a separate craft local, and was there as it shifted over to become an IBEW local. Langdell recalled that #77 "always did what it took to get the job done" at Hanford. Other craft unions were less diversified, and could not or would not perform certain tasks because of jurisdictional agreements.

The later 1960s brought changes to collective bargaining law as well. In 1967, the Washington Legislature passed RCW 41.56, providing collective bargaining for public employees in Washington State, while an effort to replace the state workers' compensation system with private insurance was defeated. A year later #77 lost its broadcast units, which were transferred to IBEW locals #73, #112, #497, #191, #46, and #76.

Big changes also came to the Northwest power grid. According to Seattle City Light's website history, "three factors began to chart new directions for Seattle City Light: unprecedented demand, environmental concern, and drought." There was a sense among many utilities that there were no obvious hydropower sites left in the region, especially as concerns over their environmental impacts increased.

The BPA and 109 of its customers formed the Joint Power Planning Council (Joint Council) in 1966. In 1968 it proposed a wave of new construction:

> The Joint Council recommended a $15 billion, 20-year Hydro-Thermal Power Program to build new thermal power plants (primarily nuclear), new dams, and new transmission lines. By 1969, the estimated cost of this effort had risen to $17.9 billion ($6.1 billion of the total would come from the federal government, and the remainder would come from participating utilities.)
> Northwest Power and Conservation Council website, John Harrison, "Bonneville Power Administration, History"

Going into the 1970s, union culture, law, jurisdiction, and even the sources of energy generation were changing. Local #77 leadership was changing too. Art Kenny lost his bid for re-election in 1971.

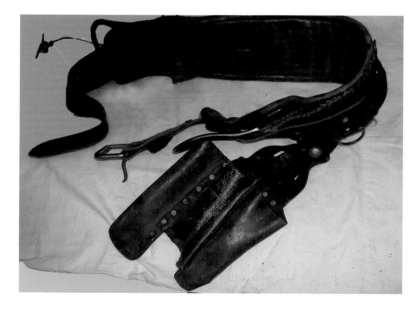

Local #77 member Ed Noyes' tool belt. Note how few tool slots there are on it.
Loren Noyes Collection.

Dad Was a Union Guy

Rick Larsen, U.S. Representative for Washington's 2nd Congressional District, provided the following remembrance of his father, a #77 lineman.

When the power goes out, you go to work. It could be your kid's birthday. It could be at the end of a 12-hour shift. It's usually in the middle of the night. But you go to work.

And so when the storm hit that early morning 20 years ago, we knew that Dad would be getting the call and heading in. The Inauguration Day storm in 1993 swept through the Puget Sound, leaving 750,000 homes in the dark. While most folks in Western Washington hunkered down, my dad and hundreds of other members of IBEW Local #77 went out into the wind and rain and began the task of restoring light, heat, and safety to us and our neighbors.

Dad was a union guy. He might not have been heavily involved in union affairs, but he sure had plenty to say about management around the dinner table. He would often say, perhaps only half-jokingly, that the linemen were always right, and management never was. I don't know if that was always fair, but his point was well taken. After all, it was the linemen like my dad who were the ones hanging off power poles in the middle of the night so that families like ours in Arlington could get the lights back on.

Dad taught me a lot about the value of an honest day's work, setting an example that I carry with me while representing Northwest Washington in Congress. I like to tell people around the District that if their kitchen light comes on when they flip the switch, it isn't luck or magic. It is thanks to hundreds of workers like my dad and those featured in this book who work 24 hours a day, 7 days a week, and 365 days a year for all of us.

Retired #77 member Loren Noyes's father, Ed Noyes, was also a lifetime #77 member. Ed Noyes (center, in leather jacket) was a career lineman for SCL who survived electrocution. Loren remembers when his father came home from the hospital he still carried an electrical charge that was strong enough to jolt anyone who touched him. Loren Noyes Collection.

Security Increased by WWP

6/6 1974

Washington Water Power Co. has built up its security patrol over the past week or so to combat increasing vandalism in the system, P. Rich Bakes, Spokane Division manager for the utility, said today.

In two separate incidents of vandalism near Davenport yesterday power was disrupted when ropes were thrown over power lines and the lines were pulled together, Bakes said. The power outages were perhaps three hours in duration, he said.

It was the third time vandals had struck the utility in as many days and Bakes said the persons responsible apparently had technical knowledge of electrical equipment.

During John Starcevich's tenure as #77's business manager, there were numerous strikes and walkouts. George Bockman, a career member of #77 as a WWP lineman, kept a scrapbook of media coverage, including these clippings from 1974. George Bockman collection.

D 6 The Seattle Times   Sunday, April 21, 1974

Negotiations to resume between City Light, union

Contract negotiations between Seattle City Light and 715 electrical workers will resume this week with the threat of another walkout hanging over the proceedings.

This time, the walkout would be a union-sanctioned strike. Workers voted 593 to 10 in ballots counted Friday night to give strike sanction.

City Light's contract with Local 77, International Brotherhood of Electrical Workers, expired March 31. Negotiations were halted when some 1,000 employes left work April 8 in a protest over discipline and managment policies of Supt. Gordon Vickery.

"The strike vote comes as no surprise," a City Light spokesman said yesterday. Local 77 members in March overwhelmingly rejected a city offer of a one-year, 8.4 per cent pay increase.

Charles Silverpale, business representative for the union, said he did not foresee an imminent threat of a strike, but intended to seek sanction for one from I. B. E. W. international officials. The King County Labor Board granted strike sanction last week to Local 77.

The strike-ballot count was made just hours after workers ended the 11-day, unsanctioned walkout by agreeing to a series of proposals for review of disciplinary procedures and City Light administration.

"Everything is settling into a familiar pattern," the utility spokesman said yesterday. "It's a welcome thing to have the work started again. We know there are problems and we intend to address those problems."

He and Silvernale both said they were optimistic that a new contract could be negotiated without a strike.

Outage Result of Sabotage

BREMERTON (UPI) — A power outage north of here yesterday was caused by sabotage, Ken Cox, local manager for Puget Sound Power and Light Co. said.

Cox said an $80,000 transformer was destroyed in the process, making it the most expensive in a recent series of acts of sabotage.

He said an oil plug had been removed from a transformer at the Chico substation and, as happened recently at Orting, all the oil ran out.

The outage started just after midnight and full power was not restored until shortly before 3 a.m. yesterday.

6/13/74

WWP Union Considering New Contract

Wage increases totaling $2.10 an hour over three years are provided for in a proposed new labor agreement being voted on today and tomorrow by striking Washington Water Power Co. employes.

More than 500 members of Local 77, International Brotherhood of Electrical Workers, are balloting in Spokane, Coeur d'Alene, Pullman and Lewiston on the proposed strike settlement reached late yesterday, a union spokesman said.

Results of the ratification vote will be announced late tomorrow or Saturday morning, said Mrs. Norman K. Turner, secretary of the local.

The proposed new agreement provides for wage increases ranging from 30 cents an hour for helpers to 75 cents for journeymen, retroactive to March 26, expiration date of the old one-year contract, she said.

On March 26, 1975, journeymen would receive an additional 70 cents an hour or the increase in the cost of living, whichever is greater, she said. The WWP employes also would receive a 10th holiday —on the second Monday in March—and the current pension plan would be adjusted to permit retirement at age 63 with no penalty.

Pay Increased

As of March 26, 1976, journeymen would receive an additional 65 cents an hour or the increase in the cost of living, whichever is greater, she said. Full retirement then could be taken at 62.

The strike, first in in the power company's history, began May 1 and involved 576 of WWP's 1,100 employes. Supervisory personnel and electrical engineers kept WWP plants operating.

JOHN L. Starcevich was elected #77's business manager in 1971. His specific role in much of what happened at #77 during his six years in office is not well documented. It was under his leadership, however, that the Central Washington PUDs came together under a collective agreement.

Bill Shaffer, who had represented #77 in Central Washington since 1956, had secured contracts with the PUDs in his region that had the best wages in the state. Those who are familiar with the PUDs in Central Washington, including John Trumble (longtime #77 business representative) and Bob Isley (a #77 executive board member from the mid-1980s until 2000), have spoken with great deference about Shaffer's ability to "work the river" (the Columbia River) in negotiations.

Part of this was Shaffer's negotiating skill. "I'd find the utility I could get some traction with, and get the terms there," Shaffer recalled. Others told how Shaffer was able to convince management they would come out best if they dealt with #77. Some of his success also came from working directly with PUD commissioners, commissioners who negotiated the contracts themselves.

Isley gave the example of Benton County PUD's Commissioners. They "all had union workers directly under them, qualified workers," he notes. "Even in apple-knocking country, where firing-at-will was a regular practice, they could see the value of the union and they didn't fight it like they do now."

Out of his individual PUD negotiations, Shaffer secured a unique collective agreement with the five Central Washington PUDs in 1972, shortly after Starcevich was elected.

Still in place, this coordinated bargaining agreement differs from a "master contract" because it is between several employers and one union local (#77), not a trade council of affiliated craft unions and locals (as in the master contracts with HAMTC or CBTC). The Central Washington PUD agreement (CWPUD) is between #77 and Chelan, Douglas, Grant, and Okanogan County PUDs.

John Trumble, the business rep who negotiated the CWPUD agreement for #77 from 1988 until his retirement in 2012, explains that the CWPUD joint agreement "has rules [for signatories] about getting together, and rules about how to get out." The CWPUD joint agreement does not fall under the jurisdiction of the National Labor Relations Board (NLRB) nor Washington State's Public Employees Collective Bargaining Law (PERC, Chapter 41.56 RCW). Trumble explains that the agreement's Unified Insurance Plan (UIP) is "what made everything jell. The UIP is the glue that holds the agreement together. It supports the coordinated bargaining."

Not long after the first CWPUD contract, Shaffer quit #77 and took a job with management within the utility industry. When he left #77, Shaffer made it clear to all he was moving on in part so he could get a pension he would be able to live on. Losing Shaffer's experience and talent triggered an effort within #77 to provide pension benefits for staffers, although it took some years. Shaffer explained how it was before the local provided a staff pension. "We [staff] get a tiny IBEW pension from the IO. When I retired there hadn't been any real increase [in this pension plan] since the 1900s. I get $67 a month for how many years did I work for IBEW, 60 years in the union?" (Bill Shaffer died in the spring of 2013.)

During the time Starcevich was #77's business manager, the energy market in the Pacific Northwest faced major restructuring. By the mid-1970s the BPA's energy projections for future production and demand showed a significant gap in capacity. In particular, the BPA told all its generating utilities they couldn't produce enough electricity to meet the PUDs' projected needs. At this point WPPSS and other Northwest utilities were "increasingly turning to thermal power plants to meet local growth," according to Hirt (*The Wired Northwest*, p. 352).

Neither the utilities nor the BPA foresaw a huge spike in oil-related energy costs. In 1973 the Organization of the Petroleum Exporting Countries (OPEC) imposed an oil embargo. Almost immediately, many of the non-hydro sources of electricity generation cost more than hydropower as their prices rose with the cost of oil. Hirt notes that this "was the first time in [either] the Northwest or the U.S. that energy prices climbed both in absolute terms and as adjusted for inflation."

As the cost of petroleum-based energy, including gasoline, soared, consumers cut way back on their energy consumption. This dropped projections for future energy demand, and reduced the projected need for more power production. The consumption decline also "challenged cost recovery for existing power plants," according to Hirt. That is, existing power plants could not expect rapid increases in consumption to pay off the already incurred cost of construction.

Meanwhile, #77 shifted its focus to Olympia. The federal Occupational Safety and Health Act (OSHA) had passed in 1970 with the aim of protecting workers from job-related death, injury, and illness. Through the Occupational Safety and Health Administration (also OSHA), OSHA required implementation of major federal safety law at the state level.

The challenge was in reconciling OSHA with existing state law. In Washington State, there was already a law those in #77 refer to "Chapter 45." Chapter 296-45 WAC, Safety Standards for Electrical Workers, was the first electrical safety law established in the United States and it served as a model for laws in other states. In 1913 #77, under then-President George L. Brooks, had drafted and led the charge to get Chapter 45 passed. Fifty-seven years later, when it came time to integrate OSHA with Washington State's worker safety laws, #77 worked hard and fast to successfully update Chapter 45 into broader and more comprehensive worker safety laws.

In 1973 the Washington Industrial Safety and Health Act (RCW 49.17) was enacted "to create, maintain, continue, and enhance the industrial safety and health program of the state, to equal or exceed OSHA standards."

While #77 was successfully working to pass legislation that would protect its members on the job, the local was also fighting implementation of affirmative action in Seattle. In August 1972 Seattle Mayor Wes Uhlman had issued an executive order establishing an affirmative action program for all city departments, including Seattle City Light (SCL).

Clara Fraser, a longtime Seattle activist, Marxist, and feminist, was hired to design the program. She decided SCL's first Electrical Trades Trainees (ETTs) would be women. After studying ETT programs elsewhere, she knew it was critical that a cohort of trainees go through the program together, rather than bringing in one worker at a time within a larger training program.

In June 1974, SCL began to implement its ETT program with a cadre of ten women who had been selected from over 300 applicants. In a departure from other ETT programs, these women became members of #77 on the day they entered the program, with their own bargaining unit. In other programs, ETTs went through a probationary period before they became union members, and then became part of their worksite unit rather than forming their own unit within the local.

Almost immediately, the ETT program ran into roadblocks. Its participants were told their training would be shortened. The ETTs' travails overlapped with a ten-day walkout by almost 1,500 SCL employees in what was called the "Coffee Break Walk-out" in the press. Most of the 80-percent female SCL clerical staffers, who were *not* represented by #77, joined in when #77's linemen and other members walked out. At issue was SCL's disciplinary code. Seattle City Light employees felt SCL's discipline code was being selectively enforced by SCL management, including SCL Superintendent Gordon Vickery.

Eleven days later, SCL employees were back at work, having secured an agreement with three major concessions, as documented by Heidi Durham, one of the ETTs:

> (1) A joint committee of employees and management representatives would write an Employee Bill of Rights and Responsibilities which would replace the disciplinary code, (2) a Public Review Committee to conduct open hearings on City Light management, and (3) a guarantee of no reprisals.
> Heidi Durham, *Radical Women in Action–the Case of Seattle City Light*, p. 3

Charlie Silvernale, #77's business manager in the early to mid-1980s, handled some of SCL's negotiations as a business rep. He cited SCL's changes in management as contributing to the problems #77 had in this 1974 walkout and in the bigger strike at SCL that followed a few months later. Silvernale explained that before Mayor Uhlman's 1972 appointment of Vickery as superintendent, SCL "was always operated by somebody who knew the utility business: J.D. Ross, Hoffman, Doc Raver [Paul Raver, former head of the BPA]. John Nelson worked his way up. Vickery had no experience."

SCL Superintendent Vickery lost whatever sympathy he had for both #77 and the ETT program shortly after the walkout, blaming the union and the ETT program for his difficulties in managing SCL.

> Gordon Vickery morphed from a publicity-loving ETT supporter into avowed ETT enemy halfway through the program. As well, the union that represented the women, the International Brotherhood of Electrical Workers (IBEW) Local 77, reacted to gender integration with periods of alternating resistance and assistance.
> Nicole Grant, *Challenging Sexism at City Light: The Electrical Trades Trainee Program*, p. 1

Local #77's contract negotiations with SCL in 1975 would eventually result in a strike. But before SCL workers went out on strike, everyone at PSP&L walked and remained out for nine weeks. C.O. Smith, #77 business rep (then and currently), remembers the situation. "Everybody at PSP&L struck, something like 1,500 people were out. Linemen, substation guys, the meter department, the call center, generation people, those in engineering. Everybody."

At issue was the size of PSP&L's line crew as well as consistent progression for B group members. Linemen crews had four to five workers, including a supervising journeyman foreman. PSP&L proposed reducing this to three to four on a crew, including a working foreman who would also supervise. In the end, PSP&L got its smaller crew size, with some restrictions on what the crew could do.

After the strike, Smith remembers the multi-year contract re-opened for wages in 1975. Local #77 negotiated a retroactive 14-percent increase in pay, "enough to buy a house off my second-year apprenticeship wages," he recalls. All of #77's members at PSP&L were rewarded for their solidarity during the strike because their wages were pegged to those of the linemen.

But the biggest 1975 strike by #77 was at SCL. John Trumble, longtime #77 business rep, remembers Ray Warren, a Central Washington business rep, had negotiated a great COLA [cost of living adjustment to wages, to compensate for inflation] for the CWPUDs. SCL workers wanted something comparable.

In Charlie Silvernale's opinion, "The City Light strike [of 1975] didn't have to happen." Seattle had what Silvernale calls a "me too" agreement, a 1951 resolution that it would pay equivalent to the average of other contracts, and this held until the 1975 strike. Silvernale recalls SCL coming to the table with a list of "comparables" that included PSP&L, some PUDs, and PG&E. The catch was that all of these contracts were effective January 1, 1975, whereas the PSP&L contract was due to expire three months later, in April.

Silvernale says there was an understanding that if everything settled, it would be possible for #77 members, not other unions, to get any pay increases retroactive to the first of the year.

It was a bad time to take on SCL, however. Mayor Uhlman was facing a potential recall. The City of Seattle Fire Department's union, IAFF #27, was challenging Uhlman over the city's affirmative action program in their department. Although #77 had supported Uhlman and was not officially involved in the recall, Uhlman lashed back at #77. Charlie Silvernale recalls Uhlman's early salvo to #77's negotiating committee. "Local #77 won't run the City. You guys can go on strike, I don't care." Silvernale relayed this message to Business Manager Starcevich who said, "Go back and bargain."

The City's offers did not meet "comparables," and #77's members voted to strike in September 1975 but didn't walk.

By October the matter was before the NLRB. The City rescinded its "comparable" resolution (#15766).

On October 17, 1975, 700 Local #77 members at SCL walked after their offer of binding arbitration was refused. Local #77 took a hard line in its January 1976 report to *The Electrical Worker*:

> This is only a very brief report of our situation. It would take a book to document all the unfair tactics we have been confronted with and are still facing. They say "You can't beat City Hall." Well, we are going to show them that whoever made that statement didn't know Local 77.
> *The Electrical Worker,* January 1976, Local #77 report

Local #77 had another problem. It had not followed through on bringing the non-union clerical workers at SCL into the local.

> IBEW Local 77 failed its promise to organize these women workers like they said they would. The women's solidarity [during the 1974 walkout] went largely un-recognized and as a result many clerical workers did not support Local 77's 1975 strike, which ended in bitter defeat.
> Nicole Grant, *Challenging Sexism at City Light: The Electrical Trades Trainee Program*, p. 4

The 1975 SCL strike ended on management's terms, with not a single union demand met after ninety-eight days on strike. Local #77 reported to *The Electrical Worker* that this contract was signed, but that "the members at all but one PUD have taken and passed strike votes."

Not mentioned in #77's report was the 1975 passage of RCW 41.58. This law created the Public Employment Relations Commission (PERC) to administer public employees' collective bargaining rights. Local #77 also did not report that as of December 1975, it had lost its outside telephone work to IBEW Local #89, a Seattle local chartered in 1967 to represent "cable television, electrical manufacturing, line clearance tree trimming, and outside telephone work."

In 1976 Hanford workers under HAMTC went out on strike. This included #77 members.

> The Hanford Atomic Metal Trades Council (HAMTC) handles the negotiations for all union members on the Hanford project. About 200 of them belong to our local.
> This also shuts down the N-Reactor plant which supplies steam for the Washington Public Power Supply System [WPSS] steam plant. Our members there are separate from the project and are still on the job catching up on the maintenance work while the reactor is shut down.
> *The Electrical Worker,* July 1976, Local #77 report

This mention of "catching up on maintenance work" at WPPSS was an understatement. WPPSS had projected a need for five WPPSS-owned 1,000-megawatt nuclear plants, three other private-utility-owned nuclear plants, and several coal plants. Plants 1 and 2 were the WPPSS nuclear plants at Hanford. All WPPSS construction was far behind schedule and way over budget as of 1973, although apparently neither problem was being reported in the press.

> At a WPPSS board meeting in 1973, Seattle City Light Superintendent Gordon Vickery expressed surprise that the projects were a year behind schedule. According to the minutes, "Mr. Vickery expressed his opinion that the public should be told this, rather than keep saying we are on schedule and then come up short."
> David Wilma, "Washington Public Power Supply System (WPPSS)," HistoryLink Online Encyclopedia of Washington State History

Things had only gotten further behind by 1977.

> A combination of management failures, a depressed economy, soaring interest rates and material costs, labor unrest, ratepayer activism and over estimation of electricity demand by forecasters was more than the effort could withstand.
> Wikipedia, "Energy Northwest"

With the 1977 elections approaching, Starcevich decided not to run again, and retired.

Setting Poles

Above: Line crew, circa 1900–1930. Local #77 Archive.
Right: PSP&L workers erecting power pole near houses.
University of Washington Libraries, Special Collections,
UW35602.
Immediately below: Peck Idaho, between 1938 and
1939. L-R: A.K. Harrison, Larry Glass, Chet Banks, Bob
Stintson, Bob Deasey with 1937 International Line Truck
in background. Local #77 Archive, Spokane office.
Bottom left: PSP&L crew raising pole with truck boom.
University of Washington Libraries, Special Collections,
UW35557.
Bottom right: WWP crew setting pole. Local #77 Archive,
Spokane office.

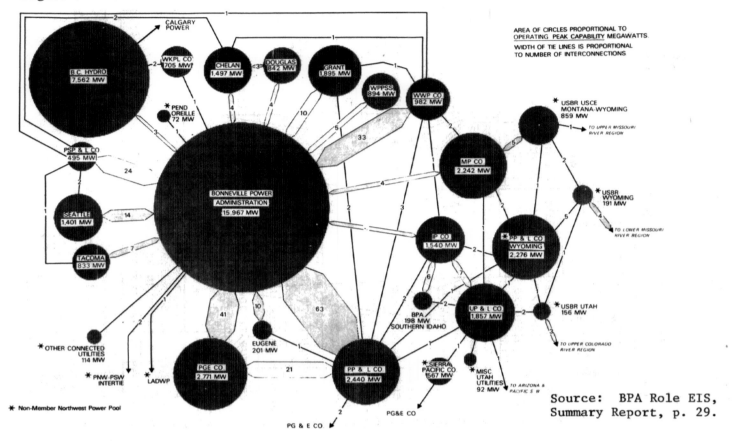

Figure 7. Northwest Power Pool Interconnections, 1977.

Source: BPA Role EIS, Summary Report, p. 29.

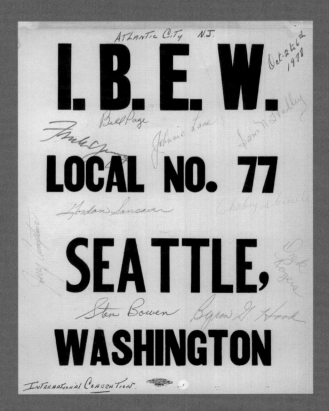

Above: Diagram of the energy system in the Pacific Northwest in 1977. Area of circles is proportional to operating peak capacity (in megawatts). Width of lines is proportional to the number of interconnections. BPA website.

Left: Local #77 poster from the 1978 IBEW International Convention in Atlantic City, signed by #77's delegates including Dick Rogers and Gordon Sansaver. Local #77 Archive.

Warren Adkins
Business Manager
June 1977–January 1978

WHEN John Starcevich retired as #77's business manager in May 1977, Bob Zehnder, a business rep, ran against another business rep, Warren Adkins, and lost. Adkins' election was contested until August 1977. Several who were on staff at the time described Adkins' sweeping appointments and firings (including that of his opponent, Zehnder) as cronyism, and several who remained on staff did not support him.

Dick Becker describes Adkins as a "good talker." (Becker worked at WWP before becoming a #77 business rep in the 1980s.) Becker saw Adkins' win as a result of membership's feeling that it was "time for a change" rather than as a result of any specific problems with the way Starcevich had run the local. Becker further describes Adkins as a poor businessman whose political views were considerably to the left of many in #77. Other staffers remember Adkins as using #77's resources to support leftist political activities that had little or nothing to do with the local.

In addition, Adkins had been an operator, not a lineman. Although #77 represents many non-linemen, linemen have almost completely dominated its leadership. This meant that as Adkins lost support from the general membership, he faced elected leaders within the local who were primarily linemen.

Adkins resigned in January 1978.

Men working in a PSP&L transmission tower. University of Washington Libraries, Special Collections, UW35601.

Richard "Dick" Rogers
Business Manager
January 1978–1981

WHEN Adkins resigned, Reuben Gable remembers that the executive board would *not* appoint Bob Zehnder to replace Adkins because Zehnder had lost the election to Adkins less than a year before. (Gable was #77's recording secretary from 1987 to 1996.) John Trumble, longtime #77 business rep, recalls that the IO was involved in Richard Rogers's appointment as business manager, the only time in Trumble's career he remembers the IO "coming in." This involvement had to do with Adkins' poor financial accountability while in office.

Once Dick Rogers became #77's business manager, he definitely took control. Dick Becker describes Rogers as "a changeable guy. He leaned on people. He was kind of a bully, especially if you presented facts he didn't agree with." (Becker served on the local's Executive Board, worked as a business rep for #77, and was #77 President from 1988 to 1990.) John Trumble also had vivid memories of Rogers.

> Rogers was bigger than me. He had long hair, and he was intimidating. He threw me against the wall one time. But he could make things happen with his bullying.... [When Rogers had become a District 9 rep], I was having a hard time with some of the other trades on a jurisdiction. Rogers showed up and made them back down. Rogers always wanted the best for #77. He was a good IO rep.

During his first year in office Rogers had two strikes to deal with. Local #77 members at the City of Richland went on strike sometime between 1977 and 1978 to try to maintain parity with the CWPUDs. Ray Warren had stepped up to represent the CWPUDs, and had done well in negotiations.

Grant County PUD workers went out sometime during this same period, again trying to match the terms of other #77 contracts, this time those in Western Washington, especially their COLA.

At about the same time, a variety of WPPSS workers were coming into #77 as members. Mike Hanson, now Avista's manager of craft training, came out from Minnesota as a licensed nuclear power plant operator to work on WPPSS Projects 1 and 2 (WNP1 and WNP2) in 1977. Hanson immediately got onto the power plant operators' negotiating committee, since he had always been a committed union activist. (Hanson served as a business rep under Rogers and was fired three different times by Rogers.)

Opposite: SCL workers taking a break, probably near the Skagit Dam. L-R: Sue Earls (ironworker), Lynne Moore (journeyman iron worker), unidentified helper, Carol Nedler (machinist). Gary Moore Collection.

Three views of iron work inside the Francis water turbine of SCL's Diablo Dam. (A picture of the Lake Chelan Hydroelectric Project generator, which has a Francis turbine, can be seen in "World War I," p 27.)

Top: Access is limited getting in and out of each "bucket," or slot, of the turbine.

Center: The view from inside a bucket as a worker swings in. The only light is what a worker brings into the turbine.

Bottom: All work inside the turbine work is performed hanging upside-down from the waist, wearing a welding hood that has an individual breathing line (seen in top two photos). All three photos from the Gary Moore Collection.

WPPSS construction brought in other new #77 members, including a wave of wiremen from the East Coast. Tom McMahon, now a business rep in the Kennewick office representing HAMTC and other Hanford workers, migrated west in 1978 as one of some thirty members of a single New York local #501 (Westchester County). He worked on WNP1 (WPPSS Nuclear Project-1), WNP2, and WNP4, literally pulling wire, miles of wire. ("Pulling wire" is also slang for inside electrical construction work.)

Sonny Trenti also came out from New York, and started working construction at WPPSS out of #112 (Kennewick construction), initially working for the contractor Fischbach & Moore. He then worked directly on WPPSS as a wireman, and did what was needed: heat-stress work with pipefitters, maintenance of temperature power, and working as foreman of a wire-pulling crew. Trenti has served #77 on negotiating teams, as a worker representative on various safety programs at Hanford, and in innovative programs to improve work quality at Hanford.

While WPPSS-related work was growing and bringing in new members, #77 was grappling with equitable representation between its Eastern and Western Washington members. To Dick Becker, who came from the Spokane area and worked at Washington Water Power (WWP), it seemed clear the eastside membership of #77 didn't have enough representation. He remembers that in 1978, with the election of Jerry Compton to #77's executive board, the board started to have regionally assigned positions. Before that, all of the board had been elected by a straight majority of the membership. Becker replaced Compton on the board while Rogers was business manager, and was eager to continue to "carry forward the eastside idea."

Local #77's increased Eastern Washington representation may have helped secure the 1980 transfer of IBEW #984's maintenance electricians into #77. (Local #984 is a Richland local, chartered in 1956, formally representing "atomic research service" workers, according to IO records.)

In 1981 Dick Rogers resigned as business manager of the local to become a Ninth District Representative for the IO. Not long before Rogers left the local he hired Don Guillot, a shop steward with WWP, as a business representative. Although Rogers officially moved on, he kept himself apprised and sometimes involved in #77 until his death in 2003.

Left: Charles Silvernale, 2011. Belew photograph.

Below right: PSP&L's Captain Energy visits the schools to promote energy conservation, circa 1981–1982. Northwest Museum of Arts and Culture/Eastern Washington State Historical Society, L2007-1.12.8.2.

Below left: PSP&L's Eastside System Operations showcasing its new Energy Management System in Redmond, Washington. University of Washington Libraries, Special Collections, UW35554.

Charles Silvernale
Business Manager
January 1981–November 1987

WHEN Dick Rogers left to work as a District 9 representative in 1981, Charles Silvernale was appointed #77's business manager, then elected the following year. Silvernale had grown up with an understanding of #77's politics. His father, H.S. Silvernale, was president of #77 from the mid-1940s to the mid-1950s. Charlie had served on the executive board in the early 1970s, worked as a business rep under John Starcevich, was fired by Adkins, and then rehired by Dick Rogers. Under Rogers, Silvernale had been involved in negotiations during the SCL strikes.

From the beginning, Silvernale had to deal with an anti-labor business climate, a regional recession, and a volatile energy market. Anti-union action came to a head almost immediately in August 1981, when President Ronald Reagan decertified a federal union after firing most of its members. Air traffic controllers had gone on strike illegally. (Other illegal strikes at about the same time had been able to secure negotiating power for their unions.) Reagan ordered the air traffic controllers back to work, calling their strike a "peril to national safety" under the 1947 Taft-Hartley Act.

Two days later, Reagan fired 11,345 members of the Professional Air Traffic Controllers Association (PATCO) and banned them from federal service for life. In what is often cited as the largest labor rally in U.S. history, some 400,000 union supporters gathered to protest in Washington, D.C. but did not change things for the fired workers.

Alan Greenspan, who served as Federal Reserve Chairman from 1987 to 2006, put Reagan's actions into context, in a 2003 speech:

> The President invoked the law that striking government employees forfeit their jobs, an action that unsettled those who cynically believed no President would ever uphold that law. President Reagan prevailed, as you know, but far more importantly his action gave weight to the legal right of private employers, previously not fully exercised, to use their own discretion to both hire and discharge workers.
> Federal Reserve website, Alan Greenspan, "The Reagan Legacy: Remarks"

Reagan made it clear that union-busting would be pursued not only by private employers, but by the federal government, to the full extent of the law.

In addition to increased union busting, utility workers and linemen faced an uncertain energy future. This uncertainty had been building for a decade. In the Pacific Northwest, projections for energy generation, supply, and demand were in flux.

Major regional droughts in 1977 had diminished the supply of hydropower. Petroleum-based energy costs had soared.

> In the late 1970s, forecasts of impending shortage led to concerns about how the limited supply of cheap hydro-generated power in the Pacific Northwest would be divided among regional utilities. Congress passed the Pacific Northwest Electric Power Planning and Conservation Act (Northwest Power Act) in 1980, which granted BPA the right to acquire new resources to serve load growth of public utilities.
> Public Power Council website, "Public Power History," 2006

Consumers had been reducing their consumption as the cost of energy rose. The public had begun to grasp the environmental cost of hydropower as well.

> In... the Northwest... the not-so secret truth is that when we turn on the lights something awful happens: in the words of Nisqually Indian leader Billy Frank, Jr., salmon come flying out.... Under the Northwest Power Act, Bonneville's administrator is required to repair the damage done to fish and wildlife of the Columbia River Basin by hydropower dams. Bonneville does this by implementing the Columbia River Basin Fish and Wildlife Program developed by the Northwest Power and Conservation Council.
> Northwest Power and Conservation Council website, John Harrison, "Bonneville Power Administration, history"

But the biggest shift in the regional power situation by the early 1980s was the public's gradual realization that WPPSS was a fiasco. The WPPSS board of directors had been slow to recognize the problem.

In 1984, board member David Lee Myers, a Wahkiakum Public Utility District Commissioner at the time, co-wrote an article that analyzed how he and others on the WPPSS board had been unable to recognize what was happening even as WPPSS became commonly known as "Whoops":

> [E]ditorials in Northwest papers began to call for investigation and governmental action. The Washington state legislature began an inquiry into the management of the projects. The majority on the WPPSS Board of Directors, while beginning to admit that some changes had to be made (like a new managing director) still felt that the plants must be built, that we would soon run out of electricity, and we had to have more.
> Elaine Myers and David Lee Myers, "Lessons from WPPSS, a $2.25 Billion Fiasco Illustrates the Drawbacks to 'Business as Usual' Approaches for Major Social Decisions"

Dan Leahy was part of the leadership of what he calls a "back-to-grass-roots ratepayers' revolt against WPPSS" that began in 1980 and created a political action committee (PAC) called Progress Under Democracy, "dedicated to putting the public back in public power." (Leahy is now a public policy consultant and organizer for Progress Under Democracy.) The organization immediately began challenging incumbent PUD commissioners who supported WPPSS.

Two years later, in 1982, Progress Under Democracy organized campaigns in fifteen counties against such commissioners. Twenty-six PUD commissioners were up for re-election; only six remained in office after the 1982 elections.

Ironically, as WPPSS went into a managerial and financial tailspin, there was a wave of rehires on WPPSS construction work. Wireman Tom McMahon had gone to California and came back in 1982 to work for a contractor on WNP2 (WPPSS Nuclear Project-2) as a "start-up," before it was actually running. Three years later, in 1985, he was hired directly by WPPSS.

Sonny Trenti had gone back to New York State after being laid off. Still on the East Coast in 1983, he put in an application at Rockwell and got a call that they wanted him to interview in Washington State the next day.

Unions had been blamed for cost overruns as WPPSS blew through every budget and deadline, but in fact the unions were simply using WPPSS's need for skilled labor to leverage good terms for contract work on WPPSS. Although lax oversight caused some workers to take advantage of the situation, it was not the journeymen of any trade who changed work orders to do and redo particular work.

To the general public, however, the high cost of union labor was simply another aspect of WPPSS's over-inflated and mismanaged finances. Even in 1982, when WPPSS was beginning to come under public scrutiny, Business Manager Silvernale knew #77 faced the public's growing disconnect with unions when the local went into any negotiation. In addition, any ability to leverage contracts through strikes had almost disappeared. "After our 1970s strikes, a sort of change came about," he recalls. "The companies had learned they could get by without us for a period of time at least."

This knowledge didn't help #77 as it went into a prolonged contract dispute with PSP&L. C.O. Smith, a second-generation PSP&L employee and a PSP&L rep at the time, remembers that there were massive layoffs in late spring 1982 that included probationary employees in all departments, union or non-union. In addition, PSP&L instituted a hiring freeze.

The apprenticeship program at PSP&L continued, but whoever topped out "had no place to go," recalls Smith. There were no new jobs. This situation wasn't covered in the contract, and PSP&L offered to allow these new journeymen to continue to work where they topped out, as long as they worked under apprentice terms and at apprentice rates of pay. Smith, as #77's business rep, said, "Make them linemen where they're at [their current work assignment]."

Because apprentices were staying on, positions that would have come up for bid by other linemen never went out. The journeymen who would have gotten these jobs, based on seniority, were angry.

The PSP&L contract would expire in March 1984. PSP&L brought in new human resources managers: Jerry Henry as director of the department and Mark Bowman as manager. PSP&L came to #77 before contract negotiations began with proposals, one on substation work rules, and one that proposed a new sub-journeyman meter tester job classification. When the negotiating team took these offers back to the membership, the line crews turned down the substation work offer while the meter department accepted theirs.

Then real negotiations began. Smith remembers that on New Year's Eve 1983, PSP&L presented their proposal. "The new guys from PSP&L had gutted the agreement," he recalls. "We did the best we could. We were on defense from the beginning."

PSP&L wanted four ten-hour work days. The IO was clear there would be nothing in any contract for more than eight-hour days, and eventually PSP&L withdrew this proposal. There was no way to negotiate for more vacation time; PSP&L was not willing to pay for any more time "not worked." Local #77's negotiating team knew what the utility companies knew. A strike would only allow PSP&L to estimate meter reading and save money doing so.

Local #77 members were furious even as they ratified the contract on April Fool's Day, 1984.

Within a year, in 1985, PSP&L came back to #77 and wanted work-rule changes to address their new energy generation system. (It would use a hydropower and combustion turbine.) The members accepted the proposal but were not happy. C.O. Smith had had enough of being a business rep and went back to the tools, bidding a PSP&L job in Redmond.

PSP&L wasn't alone in going after labor costs to help its bottom line. WWP had promoted Paul Redmond from executive vice president to president of WWP in 1982, "freeing [former president Wendell] Satre to concentrate on the mounting WPPSS crisis," according to WWP's history book, *A History of the Washington Water Power Company.*

Seattle City Light (SCL) underwent its own management changes, with five different superintendents in the fifteen years (1980–1994) following Vickery's term. In addition to the same financial pressures other regional utilities were facing, SCL was still plagued with discrimination charges and a workforce of #77 members who were largely unsupportive of affirmative action. Evidence of this is cited in *Power for the People* (SCL's corporate history) which mentions a November 1983 linemen's walkout "to protest the discipline of a supervisor accused of discriminating against a female employee."

While the PSP&L negotiations were underway, #77 had elections that brought Lou Walter in as Vice President. In 1985, Walter was elected president, and Dick Becker replaced him as vice president. That same year Idaho passed right-to-work legislation.

Silvernale was offered an IO position in California in 1988, and resigned as business manager of #77 to take it.

Above: Joe Murphy (far left, with John Horrocks, far right) at work on behalf of #77. Local #77 Archive.

Left: No one believed more in the benefits of being in a union than Murphy. An undated cartoon illustrates Murphy's attitude. The cartoon is by John Baer from The Electrical Worker, "courtesy of the American Federationist." Baer (1886–1970) was an American cartoonist and politician. IBEW IO Archive.

Joe E. Murphy
Business Manager
1987–September 1990

WHEN Charlie Silvernale moved to California for a job with District 9, Joe Murphy was appointed to replace him as business manager. It was at this point that #77 officially merged the positions of business manager and financial secretary. One individual had held both positions, however, since 1950, when Lloyd Smith was elected.

Given that Murphy was a narrowback, or wireman, it is a testament to his vision of political action that he was appointed business manager. To date, Warren Adkins is the only other non-lineman to have served in this position.

Joe Murphy lived and breathed politics. He had served as Kitsap County's Democratic chairman from 1974 to 1978, then chaired the Washington State Labor Council (WSLC). After Murphy left the WSLC, he served as chairman of the Washington State Democratic Party until 1981. From there he was elected vice president of the WSLC.

Paul Berndt, who eventually followed Murphy as leader of the Washington State Democratic Party, was an intern at the WSLC when Murphy was chair of the state party.

> He was the head of the Democratic Party at a difficult time.... The controversial Dixy Lee Ray was governor, and Jimmy Carter was president. Murphy was loyal to both at great personal cost.
> WSLC website, 2004 memorial to Joe Murphy

In the months before Murphy became business manager of #77, there had been significant changes on #77's executive board. Richard Vaughn came on as president, replacing Lou Walter, who went on staff.

Larry Duggins joined the executive board during this period, although records vary as to his actual appointment date. Duggins was inspired by Murphy, whom he considered a political guru. Murphy pointed to Idaho's 1985 right-to-work law as an example of why politics was critical to union work. "Murphy was always clear," says Duggins. "He used to say, 'You can negotiate the best contract you want, and one swipe of a legislator's pen can wipe it out.'"

Duggins took Murphy's observation to heart, and from his early days on the executive board, he served on the committee that established #77's PAC. (Duggins continued to serve on the PAC until he retired from the executive board in 2000.)

C.O. Smith described Joe Murphy as a great speaker, a "pied piper." But everybody, including Duggins, also knew that #77's pending negotiations with PSP&L were both critical and potentially perilous to #77. The local's relationship with PSP&L was "in a bad, bad place," remembers Duggins. "There was no mutual respect, and there was a lot of pressure on #77" because this was a major contract.

Smith had come back as one of #77's business reps at PSP&L. As Smith recalls, "The PSP&L nightmare continued, circa 1988. We had a great negotiating committee, very solid. Great people, and they worked together." The negotiating committee also had good representation from a variety of job classifications.

"Callouts" (work assignments in addition to regular shift work) became a key issue in negotiations. Callouts were not being taken by PSP&L's linemen. This meant management was filling in, and got paid an additional 10 percent to do so. Rick Johnson, current president of #77, was one of #77's two construction reps at the time. (Stan Jacobs was the other.) Johnson remembers that at first, he tried to keep #77 construction members from taking this PSP&L callout work as overtime, as a way of keeping pressure on PSP&L to hire more linemen. Eventually construction linemen did pick up the overtime work.

Murphy decided he would confront John Ellis, the president and CEO of PSP&L. Murphy arranged a meeting with Ellis, planning to present research he had commissioned that outlined Ellis's involvement with the Washington Business Round Table and other corporate associations promoting an anti-union agenda. Murphy let the press know about the meeting beforehand, so he would have an audience as he denounced Ellis.

When Ellis walked in and saw the situation, he walked out. Murphy had created a personal enemy.

With that, formal negotiations began. "Negotiations were horrid," remembers Smith. The membership was angry and rigid in its demands. "Four proposals were brought back to the membership that were rejected by a good margin." At this point #77's negotiating team decided to take a strike authorization vote. "Things had gone sideways," says Smith. "For members, there was no cost to voting 'no' on every proposal. A strike vote had a cost." Voting to go out on strike meant members would be without paychecks, and there was no indication a strike would provide any negotiating leverage. The membership voted against striking.

It was then that Murphy demonstrated his lack of acumen as a negotiator. He told the #77 negotiating team, including Smith and the other business representative, Bob Boode, one thing regarding a PSP&L proposal on time-and-a-half, while doing a side negotiation with PSP&L.

At the same time, Dick Rogers, a District 9 Representative (and former #77 business manager), had come to Seattle to monitor the situation. Rogers was having side conversations with John Ellis and PSP&L management. When Rogers caught wind of Murphy's dual representation of negotiations, Rogers demanded to know where the matter actually stood. Murphy's side negotiations were revealed to the negotiating committee. All hell broke loose.

Boode and Smith quit, then Murphy rehired Boode. Members showed up at the #77 offices when they realized Murphy had compromised their negotiations. When he tried to drive away they stopped him for some time. A cement mixer was involved, although no one was hurt.

The PSP&L contract was eventually signed in 1990, under the next business manager, Ray Warren, two years after negotiations had opened.

One person who knew Murphy made it clear Joe Murphy was a great political animal but a terrible financial officer. Murphy had "no business sense," as Duggins puts it. "The local was going downhill under him [financially], and he just didn't see it." Meanwhile, Murphy, in his single-minded push to make political changes, ruffled many feathers.

Murphy was the first business manager of #77 to encourage and support members' and staff's formal education in labor history and leadership. Rick Johnson attended the New School for Union Organizers at The Evergreen State College Labor Center while Murphy was in office, and feels this training changed his life. "I learned my history. And I realized organizing was what I wanted to do," Johnson recalls.

Joe Murphy resigned in September 1990. Ray Warren, appointed to replace Murphy as the local's business manager, kept Murphy on as #77's political coordinator and as the local's representative on labor councils, knowing how effective Murphy was in Olympia. Again, Murphy's political prowess was such that no one doubted his ability in this arena, even as the local had to address his failure as a business manager.

Joe Murphy, along with his twin brother, disappeared in a floatplane near Sitka, Alaska, in 2004. Their remains have never been found.

Joe Murphy, originally a wireman, was one of the few #77 business managers who did not come up out of the ranks as a lineman.

Above and left: doing bench work on WWP meters. Avista Archive.

Below: An early mobile metering unit used by WWP. Avista Archive.

Left: Ray Warren at work.
Local #77 Archive.

Right: WWP workers testing for PCB levels in transformer oil. Northwest Museum of Arts & Culture/Eastern Washington State Historical Society.

Below: WWP linemen from an earlier time, moving transformers (filled with PCBs). Many linemen recalled dipping their tools, barehanded, into used transformer oil, to lubricate the tools. Avista Archive.

Ray Warren
Business Manager
September 1990–November 1994

RAY Warren was appointed by the Executive Board to replace Joe Murphy as #77's business manager. Warren took the job knowing that #77 was in dire financial straits. Larry Duggins, who was on the board at the time, describes the situation: "Ray Warren took on being business manager of the seventh-largest local in IBEW when it almost went into receivership."

Warren took #77's books home to study and felt sick at what he saw. There was no money in #77. He called a staff meeting and explained the situation, then asked everyone to cut expenses wherever they could, starting immediately. He studied the books in more detail with Mike Stan, #77's office manager, and an accountant. Warren recalled, "I wanted to know where all the money had gone."

He soon realized that the constant acrimony with PSP&L had been bankrupting the local. Local #77 was paying all wages for any #77 union activity, and grievances at PSP&L were being filed almost nonstop. There were plenty of grievances coming in on other contracts as well. Each and every grievance committee "had free rein" to approach #77's legal counsel, racking up a huge bill.

Warren contacted Richard Robblee, #77's lawyer at the time, and explained the situation. Warren made it clear that significantly less work would be coming to Robblee from #77, and that any request for counsel would need to come directly from Warren.

Warren then explained these same terms to all of #77's staff. He called in particular upon John Trumble, a Central Washington business rep. Warren considered Trumble especially knowledgeable about and experienced in arbitration, and asked him to help other business representatives in these matters. Warren offered his own expertise as well, along with that of Dave Hutchins, who he felt "could do it [arbitration]," although Hutchins didn't have a lot of experience. (Hutchins was a business rep for #46, then worked at the Washington State Department of Labor and Industries, before becoming a business rep at #77.)

Warren eventually contacted PSP&L President John Ellis directly. He convinced Ellis that #77 had a new way of doing things and got a verbal agreement from Ellis that PSP&L would start to pick up the cost of wages for union members participating on committees, hearings, or other labor relations work. This agreement lasted as long as Warren remained at #77. (John Ellis left PSP&L in 1992, while Warren was still business manager.)

Warren refused to move to Seattle while he was business manager. He was renowned for living on a boat in Port Orchard during his terms, using his houseboat as a way of reminding himself and others that his time as business manager would be temporary.

Warren knew his way around the work, and around the local. He had come up as a lineman. He worked for BPA for seven years, topping out there in 1967, at age 35. He then continued to work out of #125 (Portland) on construction at various central Washington PUDs including the Chelan PUD, for PSP&L, taking winters off. He "put in his card" at #77 when he took a job in 1975 at Grant County PUD. Pretty soon he was unit chair, then became a member of #77's Executive Board. When Dick Rogers hired Warren as a business rep in about 1978, he worked out of the Kennewick office until he became business manager, some twelve years later.

Once appointed, Warren ran for business officer in 1991. It was a contested election. Dave Timothy (a later business manager) and eventually Bob Boode ran against Warren. Timothy and Smith together got the majority of votes, but Warren won with a plurality.

Once Warren had arrested #77's financial hemorrhaging, he instituted a shift toward "mutual gains bargaining," sometimes called interest-based bargaining. It was clear that many of #77's business representatives, as well as active members, had the attitude of "us vs. them." Warren sought out training programs in interest-based bargaining for the staff. First he brought out a woman from the IO who had been involved with the labor relations program at Cornell. Around this time, #77 sponsored a training event at Grand Coulee that included representatives from other trades who were also members of the Columbia Basin Trades Council (CBTC).

At this and/or other training sessions, Warren, at union expense, invited Bernie Flaherty, Cornell University Professor of Labor Relations, to speak. Warren encouraged both labor and management negotiators to go through this and other training. The degree of turmoil in negotiations between #77 and many employers made for a receptive audience.

Don Guillot, later business manager of #77, recalls that he personally "really got it." Guillot feels this training made him understand how negotiations could work. Both Guillot and Lou Walter (longtime business rep, former president of #77, and current business manager) traveled east to Cornell for further training.

In addition to training on less-confrontational approaches to contract negotiations, Warren repositioned various business representatives, assigning them to different contracts and regions. He replaced a number of senior business representatives because he felt they were not sufficiently vested in interest-based bargaining.

The next round of negotiations with PSP&L began in early 1994. PSP&L's press release announcing the contract indicates how much the process had changed since 1990.

> Members of the local [#77] voted on Wednesday. The "yes" vote was 62 percent of the ballots counted, with 604 voting in favor of the contract and 372 voting "no," said Mark Bowman, Puget Power's director of labor relations. The contract provides for a 3.5 percent salary increase effective Feb. 23 and an enhanced severance compensation package retroactive to Jan. 31.
>
> Both the company and the union cited a new cooperative approach to bargaining as playing an important part in the success of these negotiations. It involves both parties working together to help each other achieve their goals.
>
> "When the contract talks began, we proposed to the company that we try a new bargaining technique called Mutual Gains," said Don Guillot, business representative for IBEW Local 77. "It wasn't a perfect process, but overall it was a great improvement over the way things have been done in the past. And because the options are so thoroughly discussed, the resolutions are often much better than they otherwise would have been."
>
> Bowman said, "We believe that a partnership between the company and the union is essential to the success of our business, and we look forward to strengthening this process in the years ahead."
> The Free Library website, PSP&L press release, February 25, 1994

Warren felt he had done his job. "I was appointed during rough times. I accomplished the goals I had set for myself, and I didn't want to have to make political deals in Olympia," he said. Former business manager Joe Murphy had introduced Warren to the Olympia scene and its players. "He [Murphy] liked that part of things. I never did," Warren said. Warren also knew he had opinions that differed from prevailing views at the WSLC (Washington State Labor Council) and he didn't want to spend his time fighting with other unions or the WSLC, or in Olympia.

Warren did want Don Guillot to be his successor as business manager, so he offered Guillot a choice. Warren could appoint Guillot to the position in November 1994, or Warren would back Guillot in the next election, in 1995. Warren recommended that Guillot take the appointment, because he thought running as an incumbent would put Guillot in a stronger position. Guillot accepted Warren's offer and was appointed as business manager in November 1994.

Tree Work

Above left: Aspludh ad from the 1974 IBEW magazine,
The Electrical Worker. *At the time Aspludh was trying to expand into more diversified utility work. Tim Kelly Collection.*

Above right: PSP&L crew trimming tree branches. Note lack of helmets, upper man using hatchet, and early style of bucket (lift) truck. University of Washington Libraries, Special Collections, UW35600.

Left: Rick Johnson, current #77 president, topping a tree in 1994. Rick Johnson Collection.

This page: IBEW #77 and #125 members building the BPA transmission line between Washington and Oregon, 1939–1940. Towers were 545 feet high and over one mile apart where they spanned the Columbia River. Local #77 Archive.

Opposite page: Five of #77's business managers at the IBEW's 1996 International Convention in Philadelphia. (L–R): Dick Rogers, Charles Silvernale, Joe Murphy, Ray Warren, Don Guillot. The business manager at the time, John Horrocks, did not appear in the picture. Photo taken by Lou Walter, today's #77 business manager. Local #77 Archive.

Don Guillot
Business Manager
November 1994–June 1995

THERE is very little to document concerning Don Guillot's first, very short term as #77's business manager. He had been appointed by Ray Warren in November 1994, shortly after wrapping up negotiations with PSP&L using interest-based ("mutual gains") bargaining.

Guillot presumed he would win the election in June 1995, but he did not. In an extremely close race, he lost by 47 votes out of a potential 7,000 voters.

Unit conference during Horrocks's term as business manager. L-R: Lou Walter, former #77 president (current #77 business manager), John Horrocks (then #77's business manager), Hank Conner (former #77 business manager), Dick Rogers (Ninth District Representative at the time, and former #77 business manager), Charlie Silvernale (former #77 business manager), Don Guillot (former #77 business manager who would again serve as #77 business manager), Dick Vaughn (#77 president at the time), and Dick Becker (former (#77 president). Local #77 Archive.

PSP&L linemen wearing safety gloves while they give a safety demonstration using a miniature set of power poles. University of Washington Libraries, Special Collections, UW35572.

John Horrocks
Business Manager
July 1995–February 1998

John Horrocks was not available for interviews.

WHEN incumbent Don Guillot lost the 1995 election, he immediately hired his opponent, John Horrocks, because Guillot knew Horrocks would need time to transition into the job. Horrocks, once he took office, in turn hired Guillot as a business rep for Southwest Washington. Then, in the middle of a negotiation, Guillot got a call. Horrocks told him "he had somebody else after all, and that he was letting me go," recalls Guillot.

Horrocks had been elected because he promised the membership a "new direction and new leadership," according to John Trumble, longtime #77 business rep. Part of this new direction included firing many on staff, including several Seattle-based reps, as well as Mike Hanson from the Spokane office. Larry Rice, a business rep out of the Kennewick office, wouldn't work for Horrocks and quit. Trumble felt he was kept on staff so Horrocks would have at least one staffer with significant experience. "By the time he got to me," Trumble recalls, "Horrocks had fired so many people it made me an expert."

After he fired Guillot, Horrocks hired Steve Easton (who became an Executive Board member), as the "interim" Southwest Washington business representative. Easton worked as the region's rep for sixteen months, then told Horrocks the position was lasting too long to be interim, at which point Horrocks fired him. Easton had supported Horrocks in the election.

Horrocks hired Dave Timothy, who later served as #77's business manager, after Timothy responded to a call for new staffers. Timothy's first assignments, with no training, were the Snohomish PUD, Seattle Steam, and the Westin Hotel. "I just hit the ground running," he recalls. With little experience in negotiations, Timothy relied on office staff to assist him with contracts.

Rick Johnson was rehired as a business representative. When Guillot had narrowly lost the election, it was in large part because of a letter Horrocks sent out to the membership that Johnson had helped write. At the time, Johnson's top priority was organizing—"It was all about organizing," he remembers—and he knew Guillot did not share this passion. (As this account goes to press, Johnson is president of #77 and Guillot recently stepped down as its business manager. They have come to share a vision for #77.)

Johnson wanted to work for #77 as an organizer but Guillot had made it clear he didn't want a full-time organizer on staff. Back when Joe Murphy was business manager, Johnson had studied labor history at The Evergreen State College Labor Center. "It radicalized me. I knew we needed to organize or die," recalls Johnson.

During Guillot's brief first term, Johnson and another #77 lineman, Jack Sansaver, had proposed that #77 collect a membership fee to pay for organizing. Their idea had passed in the construction unit but #77's Executive Board voted it down. Horrocks told Johnson, if elected, he would establish a fulltime organizer position.

When Horrocks became business manager, he hired Johnson as a construction rep, not an organizer. Johnson was amenable because Horrocks assured him he would ultimately be an organizer. Instead Horrocks reassigned Johnson as a business rep, so Johnson quit. When Horrocks hired someone else as a full-time organizer, "My relationship with Horrocks soured," Johnson says.

The master contract at Hanford was in negotiation, and it was not going well. The contract being negotiated was between a trade council of affiliated craft unions, the Hanford Atomic Metal Trades Council (HAMTC), and the Washington Public Power Supply System (WPPSS). John Trumble recalls this round of negotiations with WPPSS as "collective begging." Hard times made it one of the most difficult negotiations Trumble had ever done. Dave Hutchins, another business rep with extensive experience in negotiations, had come over from Seattle to help.

Easton summarized Horrocks's infamous botching of this contract as "upside down." Horrocks went to the Tri-Cities and spoke with the negotiating committee. He then went public about the status of negotiations, saying they were "at impasse" without knowing what the term meant.

(In contract negotiations "declaring impasse" means the two sides are unable to reach an agreement and are deadlocked. Under National Labor Relations Board (NLRB) rules, the employer then has the right to impose its last, best offer. There is no requirement for mediation or arbitration once this point has been reached.)

Horrocks' declaration of impasse removed his own committee's ability to further negotiate. "We'll implement," was the WPPSS team's response when they heard what Horrocks had said. Trumble and Hutchins met with the membership and explained the situation. Trumble remembers, "I had to get membership to agree to go with whatever we could get, which we knew wasn't going to be much." The membership and the negotiating committee gave Trumble authorization to talk with management to see what he could get. Trumble, Hutchins, and WPPSS then met. Membership immediately ratified the contract and it was signed six hours before management could have implemented its earlier offer.

Sandra Polk, #77's treasurer from the early 1990s until 2002, remembers ever-growing financial improprieties by Horrocks. When the Executive Board looked into details of Horrocks's excessive phone bills, they realized he was a Nevada resident, which is not allowed by the bylaws of the local. The more the Executive Board investigated Horrocks's business practices, the more problems they found. Horrocks agreed to resign, and did so in late February 1998.

Linemen Face Fire

Right: Four pictures, taken in less than two seconds at 8:17 am on Monday, August 4, 2003, made the front page of the Jefferson County Leader. PSE lineman Rick Johnson (and current #77 president) was attempting to remove cables supporting the traffic signal after disconnecting electrical power. The cables were jerked out of his hands as the corner utility pole was snapped by a falling wall. He was not injured. Photo by Barney Burke, for the Jefferson County Leader.

Above: A WWP lineman using a firehose from a power pole to fight a 1969 fire in a Davenport, Washington fire. Note hose is wrapped around the pole. A second lineman, lower on the pole, is assisting. Northwest Museum of Arts & Culture/Eastern Washington State Historical Society, L2007-1.16.24.1.

Opposite, L-R: A City of Richland crew (Larry Pitman, Jerry Hexum) was called on an outage on a two-pot, or open delta, bank. They took the gate (fuse holder) out of the cutout and re-fused it. This is what happened when they closed it in. Crew foreman Terry Gonce was leaving the area and wanted some pictures of the crew. He triggered the camera just as it happened. Local #77 Archive.

113

Governmental decisions are being made which will affect your electric service for decades to come.

Voice your concerns!

YOU CAN MAKE THE DIFFERENCE!

Will Deregulation Short-Circuit North America's Electric Power Supply?

How the coming deregulation of the electric power industry may impact consumers, utility workers, businesses and investors

Above: Deregulation of the utility industry was a major factor in PSE's corporate restructuring. As part of this, PSE shifted much of its in-house linework to Potelco while Dave Timothy was #77's business manager. Larry Duggins Collection.
Left: Dave Timothy is part of a three-generation tradition of IBEW linemen. Dave Timothy Collection.

Below: Jason and Don Timothy, Dave's son and father. Dave Timothy Collection.

Dave Timothy
Business Manager
February 1998–June 2004

IN February 1998, Dave Timothy was appointed business manager to replace Horrocks. Three months after his appointment, Timothy was up for election and won. Timothy remembers that the local's finances were in terrible shape when he came on: "We were in the red." He studied the budget to see where he could cut expenses. He saw lots of calls to attorneys and lots of attorney fees, much like when Ray Warren came on after Joe Murphy.

There were a lot of grievances that were not moving forward in a timely manner. Timothy was concerned that extended delays might jeopardize #77's position under the "duty to provide fair representation" (DFR, under the National Labor Relations Act). Timothy decided to put together a grievance team that would meet before going into any arbitration that involved #77, no matter what.

The team consisted of then-business rep Patrick Lorang (or Lorraine) and Dave Timothy, with a possible additional business rep who could also replace Timothy on the team. The shop steward and the assigned representative presented their potential arbitration case to this team; the team then decided what action to take. Timothy prided himself on #77 never losing a grievance under his team system.

Timothy also modernized the local's business procedures. "It wasn't frivolous stuff," he recalls. "When I got my first check as business manager it was handwritten. It didn't have SSI information, or any of the deductions on it."

But Timothy had a much bigger problem to deal with. The entire utility industry was undergoing deregulation. The federal government had launched these changes starting in 1992.

> In one of his last acts as President, George [H.W.] Bush set the wheels of energy deregulation in motion nationwide when he approved the Energy Policy Act (1992). It aimed to break utility power monopolies across the country by opening up control over the transmission lines that deliver power, while effectively deregulating the price of wholesale electricity. But the Federal Energy Regulatory Commission [FERC] left the details of how to deregulate the retail side up to the individual states.
> Anthony York, "The Deregulation Debacle," *Salon*, January 30, 2001

In 1996, California was the first state to deregulate its energy market (California Deregulation Plan, AB 1890), followed by New Hampshire and Rhode Island. That same year, Washington Water Power (WWP) gained FERC approval to market wholesale electric power nationally, and formed a subsidiary, WWP Resource Services, to do so. From then on, WWP focused on expanding nationally.

By the end of 1996, WWP planned to begin brokering electricity in the southeastern United States, the first step toward developing into a utility involved in markets throughout North America.

> In the wake of the approval from the FERC, a new alignment of companies emerged that gave WWP a national profile. In 1997, the company formed Avista Energy, a nationally oriented energy trading and marketing subsidiary.
> Funding Universe website, "Avista Corporation History"

(WWP would change its name to Avista as of January 1, 1999.) The Enron Corporation bought Portland General Electric (PGE) the same year, which would eventually lead to PGE's bankruptcy.

John Trumble would have occasion to see what deregulation meant when he went to Montana in 1997 to bring workers at the U.S. Bureau of Reclamation (USBR) Hungry Horse Dam into #77. The USBR workers were in IBEW #283, a Boise "outside" local. It wasn't a good match. The workers wanted to get out of #283, and #283 didn't want to represent them. When Hungry Horse Dam became "remote" (automated) and therefore separate from Grand Coulee Dam operationally, #77 conducted an election and won. The IO then changed the unit over to #77.

By the time #77 signed its first contract at Hungry Horse, in November 1998, Montana Power (MPC) had sold its power plants to out-of-state utility companies Pennsylvania Power and Light and Northwestern Energy.

> MPC was built with money from citizens' pockets as a public utility with guaranteed profits of 13%. The whole snafu started when the 1997 legislature passed Senate Bill 390, the so-called "Energy Deregulation Bill." This was a blunder of historic proportions; it allowed the company to sell its subsidiaries for billions [$2.7 billion] of dollars.
> Montana River Action website, "The Failure Of Montana Power Company"

This incident didn't directly affect #77 or its members, but other corporate restructuring did.

In February 1997, Puget Sound Energy (PSE) was created when PSP&L acquired Washington Energy Company, a natural gas utility. According to Funding Universe's PSE company profile, "After paying over $36 million in after-tax charges, the company's [PSE's] net income dropped approximately 60 percent." Almost immediately, PSE put in its first request to the Washington Utilities and Transportation Commission (WUTC) for a blanket rate increase. Part of the justification PSE gave was that it had inherited collectively bargained "restrictions" that were driving up the cost of operations.

Dave Hutchins and Art Locken were interviewed together for this book. Hutchins was a business rep for #77, then worked at the Washington State Department of Labor and Industries. Locken served on the #77 Executive Board from 2001 to 2006, and started IBEW's Caucus at the National Safety Council. Both Hutchins and Locken point out that the PSE merger wasn't simply a result of PSE corporate culture, it was driven by activities on Wall Street and in the entire utility industry.

Locken summarizes, "Coming on, Dave Timothy couldn't have stopped a lot of the changes that came with the creation of PSE even if he'd had a lot more experience." Timothy did however, have hands-on experience with PSE in that he had been PSP&L's business representative before he was appointed business manager.

The first PSE-related issue that #77 had to deal with was the union affiliation of those workers who came to PSE from Washington Energy Company (WEC). None of them had been in the IBEW; most were members of other craft unions.

Charges of raiding, i.e., over-stepping its jurisdiction, were filed against #77 by other craft unions. The case quickly went to the National Labor Relations Board and Timothy went to Washington, D.C. to work with the IO on the case. In the end #77 was found to have jurisdiction except for the plumbers and pipefitters. The new #77 members were "good union workers," remembers Timothy. But no matter how good they were, many #77 members who had been with PSP&L didn't want to dovetail their seniority with these new members.

At some point after he became business manager and before the PSE merger, Timothy studied in Maryland at the AFL-CIO's National Labor College (formerly the George Meany Center for Labor Studies). Timothy remembers explaining #77's challenges to an instructor who told him he would have to decide on one of three approaches in contract negotiation.

He could "give members what they needed," except that different members needed different things. He could unify the membership on a single issue, which was almost impossible. Or he could "dice 'em [the membership] up and get B group contracts" for portions of the membership. That is, he could break with #77 tradition and let non-linemen job classifications settle for proportionally less than the linemen. Timothy decided to negotiate different terms for different job classifications.

As #77 began its first negotiation with PSE, the history of bad behavior from both #77 and PSP&L repeated itself. As part of its restructuring, PSE had changed its management personnel, so #77 was dealing with different individuals. "They got rid of the last of their hard-core labor-oriented guys, the ones that would personally stand by the contract," recalled Hutchins.

One major and continuing issue was overtime for linemen. Work hours had expanded to a point where more than 55 percent of the wages paid to PSE linemen wages were in overtime (OT). PSE asserted that linemen were creating much of their own OT, but the linemen at PSE could see there was a significant shortage of linemen for the volume of work. John Cunningham, longtime #77 business rep, points out that "Everybody *knew* PSE was going to make changes, big changes."

Steve Koreiva, a #77 lineman, remembers that National Energy Contractors Association (NECA) contractors, using #77's construction unit members, were already on PSE grounds picking up the OT. (Koreiva was working for SCL at the time, and has worked in construction for a good part of his career.) The NECA contractors (Superior Electric, for example) had good equipment, unlike PSE.

Gary B. Swofford, PSE's senior vice president and COO, invited Timothy to lunch and dropped a bomb. PSE wanted to outsource "everything" using a "service provider," although it didn't yet know who the service provider would be. Timothy immediately spread the word to the #77 staff, the Executive Board, the membership, the IO, and the public through a series of press releases similar to what was sent to *The Electrical Worker.*

> WA-Puget Sound Energy (PSE) announced a radical restructuring plan that would eliminate more than 1,200 union jobs. The work currently done by our members would be transferred to contract workers. PSE has talked about offering a transition program that would allow existing PSE workers to be hired by contractors to do the work that they now do as PSE employees.
>
> Our local union officers and staff are doing everything they can to protect the interest of our members. To the best of our knowledge, this is the most massive elimination of jobs any utility has attempted as part of restructuring. There is no template to measure the impact to the employees and their families, nor to the public regarding customer service and system reliability.
>
> Problems at PSE have affected the construction work picture. However, at this time, there are still jobs available for journeyman linemen at several utilities, as well as work for journeyman tree trimmers. Please call 206-323-0585 or visit our Web site at www.IBEW77.com for an updated work picture and information on the PSE restructuring.
>
> Sherman Williams, Jr., P.S. [Press Secretary]
> *The Electrical Worker,* May 2000, #77 report

PSE hadn't settled on the details, but PSE *was* going to outsource work. While PSE decided how it would restructure its line work, #77's construction workers didn't mind getting all the OT they could, working for various contractors. Potelco, Inc. was one of these contractors, although it was small and had poor equipment relative to many of the others. It did, however, change its corporate structure.

> In 1998, Potelco and three other contractors merged together to create a network of highly respected and experienced contracting firms from many disciplines across the nation, thereby creating an IPO on the NYSE and forming Quanta Services.
>
> Potelco website, "Welcome to our site"

The real question was which contracting company would get PSE's linework. In the meantime, PSE had created a third employee category of #77 linemen who worked directly for PSE. They worked within a new and separate department for in-house construction, with separate project centers. John Cunningham worked in this department at the time, and remembers it was clear that part of its purpose was to quantify how much work would fall into this category.

PSE put out a request for proposals (RFP) to contractors nationwide and between twenty-three and twenty-seven contractors bid, including Enron. PSE incorporated aspects of all the contractors' proposals into a retooled RFP, then rebid it to twelve of the original bidders. There was some variability in PSE's proposals as to what work would be outsourced, at least at first. PSE rebid again to six contractors, then three, including Potelco and Henkels & McCoy.

Timothy was meeting with a Henkels & McCoy representative when he got a call from PSE letting him know PSE had awarded Potelco all of the outsourced work. Potelco was already a party to #77's NECA agreement as a construction contractor, so at least #77 would have jurisdiction.

PSE and Potelco completed negotiations on the terms of their contract while Local #77 had to negotiate the terms of layoff and pension for the "downsizing." Timothy remembers that #77 negotiated as though PSE were closing a factory, although unlike a factory, a utility cannot simply stop and then restart.

PSE shifted all its linemen's work to Potelco, along with its engineers and "drafters" (those who prepare technical documents under the direction of an engineer). Warehousers, former #77 members, were not covered in #77's NECA agreement, and did not sign with #77 until some years later.

"After all the talk," Timothy recalls, "PSE did *not* outsource its substations, servicemen, those who worked on generation, or 'troublemen' [repairmen]." Its customer service representatives (CSRs) stayed in #77 as a unit, with some of PSE's call centers shut down but its big call center in Bellevue remaining open.

The final "quit package" (settlement for employees who voluntarily retired), drafted by #77 rather than PSE, was calculated using years of service, age, and other factors. Membership voted to accept the contract.

On a Friday, many at PSE went home without a job. The following Monday, most returned to the same worksite and began doing the same work, for Potelco. For PSE's former linemen, it was more complicated. They queued up at the #77 hiring hall that Monday in a line that went out the back of the hall, according to dispatchers. Many who had been BA members needed to become A members. (BA members do not participate in the IBEW's Pension Benefit Fund; A members do.)

There were other changes at #77. One was a major dues increase, in late 2003 or early 2004, partially to cover legal expenses incurred in negotiating with PSE and Potelco. According to many, Timothy became distant from members of the Executive Board, former supporters, and some staff during the PSE-Potelco transition. As the 2004 election approached, he lacked support from these individuals.

Don Guillot ran against Timothy, and won. Timothy returned to the tools, where he has a reputation as an excellent lineman and foreman.

Local #77 construction crew assembling towers. Pre-assembled sections are lifted into place while crew bolts a tower together. Local #77 Archive.

INTERNATIONAL
BROTHERHOOD OF
ELECTRICAL
WORKERS

1125 Fifteenth Street, N.W.
Washington, DC 20005
(202) 833-7000

J. J. BARRY
International President

JACK F. MOORE
International Secretary

IBEW 100
1891·1991
A CENTURY OF SERVICE

SENT VIA FAX

January 24, 1992

Mr. Ray D. Warren
Business Manager
Local Union No. 77, IBEW
P. O. Box 12129, Broadway Station
Seattle, WA 98102

Dear Brother Warren:

In response to your January 16, 1992 letter of request,
Local 77's Political Action Committee is hereby authorized
to reproduce the IBEW Logo on its PAC brochure, subject to
the following requirements:

1) The L.U. 77 PAC must be referenced
   with the IBEW Logo.

2) The brochure must be made by employees
   of an AFL-CIO affiliated union, including
   the printing or other method of repro-
   ducing our Logo.

Best wishes.

Fraternally yours,

Jack F. Moore
International Secretary

JFM:sta

Left: January 24, 1992 letter to #77 from the IBEW IO authorizing #77's Political Action Committee (PAC). Larry Duggins Archive. Above: PAC contributors traditionally receive caps for each year they contribute. Local #77 Archive.

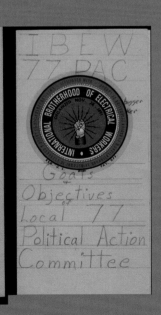

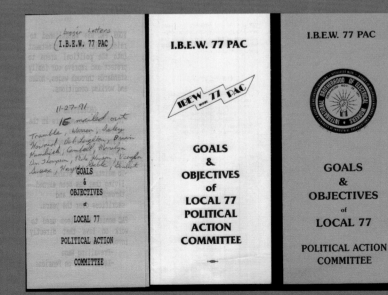

Various drafts of #77's first PAC brochure to members. Handwritten note by Larry Duggins (longtime member of #77's executive board) dated November 27, 1991 notes "Sixteen [copies] mailed out [to John] Trumble, [Ray] Warren, [Bob] Isley, Howard, Bob Leighton, [David] Humlicek, [Steve] Campbell, Marilyn [Davis], Don Thompson, Mike Hanson, [Dick] Vaughn, [Alan] Sussex, [Thomas] Hayes, [Reuben] Gable, [Don] Guillot." All named have long histories of activism within #77. Larry Duggins Archive.

Don Guillot
Business Manager
July 2004–July 2013

WHEN Dave Timothy began his re-election campaign in 2004, he made it clear he was not running on a slate with those Executive Board candidates who were also up for election. Don Guillot was then recruited to run by a group of coworkers. Guillot remembers, "Their thinking was, I had the pedigree [had been business manager and business rep]." When he decided to run, he immediately resigned as the Southwest Washington business representative, which meant he was unemployed. He went back to work at Metro Transit (King County), picking up shifts.

This time Guillot ran to win, knowing he had the support of many of #77's "greybeards" (longtime workers). He went after every vote, making sure each member, in every job classification, understood how Guillot would represent every member and every unit.

Guillot was elected in 2004 and has continued to serve as #77's business manager for nine years. He is perhaps the only #77 business manager who has won two elections as an uncontested candidate. Guillot did not seek re-election, and his term ended in July 2013.

Don Guillot's inimitable style as reflected in a sign he kept in his office while business manager. Photo by Kellie McGuire, 2013.

121

PHOTO BY JJ Kneisle

The image at left is a detail of the scrapbook photo below, which shows a joint meeting of striking Seattle and Tacoma workers on November 14, 1917. J.J Kneisle photo. Local #77 Archive.

The sign at the middle of the back reads:
> Telephone Employes [sic] Locals 42 & 77 IBEW 1500 strong

The torn-off sign at the right reads:
> We are 1,000.

Picket signs read, from left to right:
> Would you want your daughter try [sic] to live on this?

> We cant [sic] trust the telephone company.
> It does not keep its word

> If the gov'nt takes over the tel. co.- we will be back to work

> Help us be patriotic. Make the company be as patriotic as we are

> This is what we should get: Students for 9 hours' $1.50 per day
>   Exp. operators $2.50- $2.75 per day

> This is what we are getting: Students for 9 hours' $1.10 per day
>   Exp. operators $1.60- $1.80 per day

> Equal rights to all: Women have the same right to organize as men

> The IBEW bought 5000,000 $ of liberty bonds and postponed their strike. What did the tel. co do?

JOINT MEETING OF SEATTLE/TACOMA STRIKERS 11-14-17.

Lou Walter
Business Manager
July 2013–current

LOUIS R. Walter ran unopposed and was elected business manager in July 2013, as this book went to press. A longtime business rep, Walter has served on #77's executive board as treasurer, vice president, and president.

Right: Lou Walter, circa 1998. Local #77 Archive.

Training and Apprenticeships

The lineman's craft was traditionally passed on through hands-on experience, supervised by journeymen linemen. One's work attitude was as important as one's aptitude for the skills of the trade. A worker would begin as a groundsman, learn the tools, and one day, when a crew was short, would be called upon to start climbing. In 1900, #73 (Spokane) went on record against apprentices.

> Companies of all kinds around Spokane want men, but good men. We absolutely refuse to have anything to do with apprentices, as they have been a hold back to us long enough.
> *The Electrical Worker,* September 2, 1900, #73 report

Twenty-nine years later, the same local was systematizing the way "helpers" came into the union, recording in their minutes of November 25, 1929, "M&S [moved and seconded] that all helpers who have been working on permits for one year or more shall be eligible to join L.U. 73... carried."

In 1937, the federal government enacted the National Apprenticeship Law (50 Stat. 664; 29 U.S.C. 50), also known as the Fitzgerald Act,

> to promote the furtherance of labor standards of apprenticeship...to extend the application of such standards by encouraging the inclusion thereof in contracts of apprenticeship, to bring together employers and labor for the formulation of programs of apprenticeship, to cooperate with State agencies in the formulation of standards of apprenticeship.

Washington State passed its own certification program in 1941, Apprenticeship (RCW Chapter 49.04). Currently, the Department of Labor and Industries (L&I) certifies linemen apprenticeship programs within the state.

By October 1945, shortly after the end of World War II, #73 had a more formal program in place, although it is unclear if it was for linemen.

> Our apprenticeship plan in conjunction with the returned veterans and the electrical contractors is going forward in very successful manner, notwithstanding the fact that our members in the armed forces must be considered.
> *The Electrical Worker,* October 1945, #73 report

Documentation of Local #77's participation in various certified apprenticeship lineman programs begins in the early to mid-1950s. Individual employer apprenticeship programs seem to have come first. George Bockman, a career Washington Water Power (WWP) lineman, topped out at WWP in 1950. Seattle City Light (SCL) also created an apprenticeship program for city employees.

> This was set up by establishing an overall "Joint Advisory Apprenticeship Committee" with separate subcommittees for each craft. H. S. "Hi" Silvernale, our past president, served on the Joint Advisory Committee from its formation until he retired.
> *The Electrical Worker,* May 1965, #77 report

Prior to formal apprenticeship programs, IBEW correspondence and minutes from various regional locals refer to "helpers."

Employers also use trainee programs. Historically, a trainee program sometimes precluded a formal apprenticeship. Local #77 business reps John Trumble and John Cunningham worked their way toward journeyman status as PGE trainees in California, which had adopted a Joint Apprenticeship Training Committee (JATC) program in 1975. Trainee programs include some mix of on-the-job training and required classes, sometimes with optional or additional community college classwork.

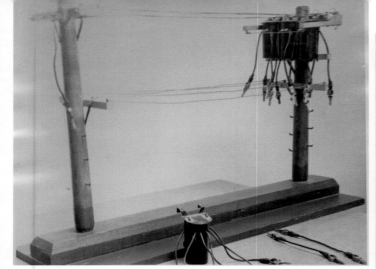

Above: Desktop model of power poles with transformers. Local #77 Archive, gift from Frank Baker.

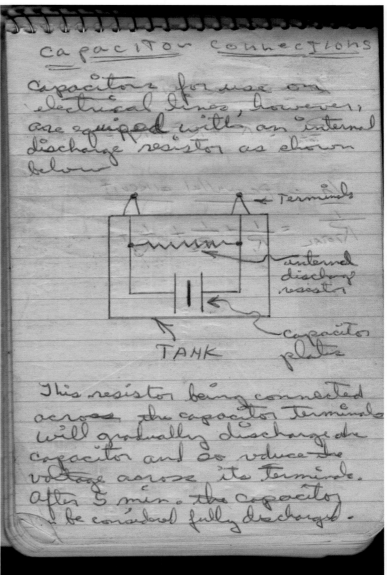

Capacitor Connections

Capacitors for use on electrical lines, however, are equiped with an internal discharge resistor as shown below.

← Terminals

internal discharge resistor

TANK

Capacitor plates

This resistor being connected across the capacitor terminals will gradually discharge the capacitor and so reduce the voltage across its terminals. After 5 min. the capacitor be considered fully discharged.

Apprenticeship class, circa 1930s. IBEW IO Archive.

Above right: Page from journeyman lineman Dick Shelley's notes as an apprentice. Dick Shelley Collection.
Below: The 2007–2008 Snohomish PUD first-year apprenticeship class. L-R top row: Jacob Larson, Steve Ingrum, Pat Kelly, Reid Maki, Jeff Finch, Lijah Manus, Phillip Weller, Torre Olson. Below: Instructors Carlos Tostado, Jeff Roberts. Timm Kellly Collection.

Form 702   T.W.W.P. Co.

APPRENTICESHIP AGREEMENT

EMPLOYED IN WASH

THIS AGREEMENT, entered into this __22nd__ day of __April__, 19 __49__, between THE WASHINGTON WATER POWER COMPANY, hereinafter referred to as the EMPLOYER, and

__George H. Bockman__, born __March 29, 1924__, hereinafter referred
(Name of apprentice)   (Month) (Day) (Year)

to as the APPRENTICE (and if a minor) _____ hereinafter
                                                            (Name of parent or guardian)

referred to as his GUARDIAN.

WITNESSETH THAT: WHEREAS, in order to preserve and perpetuate the skills essential to true crafts-manship and to maintain the ranks of skilled mechanics; and

WHEREAS, Apprenticeship Standards have been developed in conformity with the standards recommended by the Federal Committee on Apprenticeship, Bureau of Apprenticeship, U. S. Department of Labor; and

WHEREAS, the APPRENTICE through his GUARDIAN has expressed a desire to enter the required period of apprenticeship, subject to the aforementioned Standards,

NOW, THEREFORE, in consideration of the premises and the mutual covenants herein contained the parties hereto do hereby agree as follows:

THAT, the EMPLOYER shall provide employment and training to the APPRENTICE in the trade of

__Lineman__, in accordance with and under the terms contained in the Standards herein referred to.

THAT, the APPRENTICE shall perform diligently and faithfully the work of said trade during the period of apprenticeship, in conformity with the aforementioned Standards and in accordance with the rules and regulations of the Joint Apprenticeship Committee established by said Standards.

THAT, either party may terminate the agreement by written noticification to the other party. Provided, that if notification is submitted after the completion of the probationary period set forth in the Apprentice-ship Standards, the reasons for termination shall be given.

THAT, the GUADIAN will make all reasonable efforts to assure proper and diligent performance by the APPRENTICE of all obligations assumed under this agreement.

THAT, the apprenticeship term began on the __22nd__ day of __April__, 19 __49__

Credit of __17 months__ has been allowed on the term of apprenticeship.
        (Hours or months)

The Apprenticeship Standards referred to herein are incorporated in and made a part of this agreement. In witness whereof, the parties hereunto set their hands and seals:

THE WASHINGTON WATER POWER COMPANY   Registered: JOINT APPRENTICESHIP COMMITTEE

By _____   By _____ Date AUG 4 '49
                                                       Secretary
George H. Bockman
        Apprentice   FEDERAL COMMITTEE ON APPRENTICESHIP

Address __Pullman, Washington__   By _W. F. Patterson_ Date MAR 10 1950
                                                       DIRECTOR
Guardian_____

EMPLOYED IN WASH

George Bockman's JATC-approved apprenticeship agreement of April 29, 1949. This allowed him to begin his apprenticeship with WWP. George Bockman Collection.

Below: Apprentices practicing pole work, circa 1955 (unknown location). Local #77 Archive, gift of Clem Ceeber.

Opposite: The first linemen's apprenticeship class sponsored by #77 and the JATC in 1953. Back row from left: Dave Kilnman (#77 business rep), Irving Pattee (in tie and dark leather jacket, a #77 lineman, possibly on #77's executive board at this time), Bart Lindsey (center, in white hat), Vern Fitch (second from right in coveralls), Bill Cassils (possibly Kassils). Front row: Harold Fitch (far left, Vern Fitch's brother), Gene Boxx (in Stetson, second from right), Bill Lund (far right). (Gene Boxx and Bill Lund both helped identify others in their class.) Local #77 Archive.
Gene Boxx remembers he and Bill Lund were working for Montgomery Electric on the Umatilla-McNary Dam-Maupin line at the time. Both Boxx and Lund remember driving to Seattle as soon as they got off work on a Friday, staying where they could in Seattle, and driving at breakneck speed back to work Sunday afternoons. Bill Cassils (possibly Kassils) was also working for Montgomery Electric.

As competition to get into a certified lineman apprenticeship increases, line construction schools have begun to provide pre-apprenticeship training. Since 1992, Avista and Spokane Community College (SCC) have cooperatively run the Avista/SCC Lineworker School, a pre-apprenticeship program that uses Avista training facilities.

Currently SCL, Avista, and the Snohomish and Douglas County PUDs operate their own lineman apprenticeship programs. Many smaller PUDs and REAs work with the National Joint Apprenticeship and Training Committee (NJATC), maintaining their own standards and fieldwork requirements, but arranging classwork through NJATC.

Construction lineman apprenticeships have a slightly different history. Bill Stone, current Training Director with the NW Line JATC, provided a letter from the Northwest Line Constructors Chapter of NECA that notified its members of an "Apprenticeship and Training and Upgrading Program for linemen" that began operation in February 1953, "upgrading construction linemen to journeymen." The first class of thirty had completed classwork and related homework, and began on-the-job training shortly after this letter went out.

The actual NECA-Four Local Agreements do not include explicit articles outlining an apprenticeship program until November 1957, although there are references to apprenticeship rates of pay in earlier agreements. ("NECA-Four Local Agreement" refers to the agreement between the Northwest Line Constructors Chapter of the National Electrical Contractors Association (NECA) and four regional IBEW linemen's or construction locals, #77, #125, #483, and #659.)

On November 25, 1957 the NECA-Four Local Agreement was signed, and had added an "Apprenticeship and Training" section (Article VIII). The agreement created a Local JATC, to "make local rules and requirements governing the selection, qualifications, education, and training of all apprentices" which would then be submitted to NECA and the four locals. The agreement also committed all employers to pay half of one percent of their gross labor payroll into a fund established for training and apprenticeship, and gave the JATC power to set the ratio of apprentices to journeymen, as well as power over transferring apprentices between jobs or shops.

Ray Fichtner, a retied member of #77 and an IBEW journeyman lineman for forty-seven years, remembered taking advantage of "open enrollment" in #77's apprenticeship program. Age-limits on entering the program were temporarily dropped and he commuted from Idaho to the Seattle area to take advantage of the training.

Other linemen of his era and slightly before had various reactions to ever-more formalized apprenticeship programs. Research for this book revealed a number of second- and third-hand accounts or references to a wildcat strike in the early 1960s. No specific documentation was found. Some versions attribute the wildcat strike and/or the attempt to secede from #77 to a backlash against stepped-up apprenticeship requirements for linemen.

More recently, in 1974, women entering SCL's pre-apprenticeship training program, as Electrical Trades Trainees (ETTs) faced significant administrative obstacles and on-the-job harassment. *(See "A Bumpy Road for Tradeswomen," page 130.)*

"All Hung Up"
Jerry Smoot, a retired lineman and foreman with Puget Sound Energy back into the days when it was Puget Sound Power & Light Company, provided this story.

In 1977 while working for Puget Sound Power & Light at the Burlington Line headquarters in Skagit County as a lineman, I asked fellow lineman Dick Treptow if he was interested in competing in the Governor's Washington State "Pole Top Rescue" contest, and he said yes. This involved rescuing a 200-pound mannequin from the top of a 40-foot pole while in contact with hot wires. We realized we needed a third person to be the dummy up on the pole that needed to be rescued so we got an apprentice lineman by the last name of Frombeck (I can't remember his first name and I'm not sure of the correct spelling of his last name, either). We went to the state contest, which made the "PSE Safety" department really happy.

We came in dead last out of four teams but we competed. The next year Dick and I both became CPR instructors and we set up a new pole to practice on for the next year, too. We did not have our third person to be the dummy so Dick and I built a 200-pound mannequin out of an alder tree. We used 5/8″ bolts to put the body together, with lineman leather gloves full of concrete, and old lineman boots full of concrete. Then we used rope around the body to give it body size, had a square block of wood for the head, took the head strap out of a hard hat and screwed the hard hat to the head. For the finishing touch we used a Halloween mask for the face. This was one tough-looking lineman mannequin, for sure! We also used a pair of coveralls from shop mechanic Paul Erickson. His name was on the coveralls, so we called our dummy "Paul" too.

We did not have access to a bucket truck because we only had one at the headquarters and Bob Bean (our superintendent) said this would stay with the high line crew. So we used a set of four-inch blocks to raise this darn thing up the pole and get it right into the wires, as per state guidelines for the upcoming contest. It got to be such a problem getting this darn mannequin up without a bucket truck we finally talked Dick Keys (the apprentice) into being the dummy.

We left our wonderful mannequin on the truck dock out of the weather. After about a week of practice with Dick Keys as our dummy we came into the shop to find our mannequin hanging from the I-beam over the line trucks in a hangman's noose. We went into the shop and Bob Bean came up to me and said he thought it was funny that we hanged our mannequin. I told Bob someone else had done it as a joke and Bob kept laughing. We went home. The next morning Bob Bean was ready to fire me and Dick! Someone had called the Skagit

County Sheriff and the State Patrol at 3 am saying someone had hanged themselves at the Burlington Line Headquarters! The truck bay lights were left on and from the freeway you could clearly see a body hanging.

It was on the front page of the Skagit Valley paper, too. Boy, oh boy, were we ever in the news. Bob Bean was so upset, he started yelling at me at 7:45 am to get that damn thing out of his shop, so I got "Paul" out of the shop lickety-split and put him into my pickup truck with the seat belt to hold him in place while I drove back home to Bellingham that night.

I am proud to say that Dick Treptow and I finished in first place for the next three years straight. In our last year we spent sixty-five hours practicing "Pole Top and CPR" together! We even had to get a draw knife to shave the pole after so many gaff marks from my climbing gaffs. I don't believe that any other utility team has ever matched our success as a Pole Top rescue team.

I was told years later that the people responsible for hanging our mannequin were Bob Shepard, Jerry Gilbertson, and Bill Eastman. I would not put it past them either, but they never fessed up.

This page and opposite page: Apprentices practicing a pole rescue with a living "victim", circa 1953. The circuit is energized, with lines only partially covered. No one is carrying a rubber glove bag, so probably none were in use. Multi-circuit "traps," or crossbars, would now be combined for better access. Local #77 Archive, gift of Frank Bowen.

A Bumpy Road for Tradeswomen

The 1974–1976 Electrical Trades Trainees (ETT) program at Seattle City Light (SCL) is documented here as an example of the difficulties facing women and minorities coming into the trades, and into #77 in particular. In the aftermath of these events, the local became more open to journey-level women, and some of the individual women involved have become career and journey-level utility workers.

Prior to the ETT program, another Washington State IBEW local fought a significant battle over affirmative action. In June 1970, Federal District Court Judge William Lindberg ruled the apprenticeship programs and hiring practices of IBEW #46, Seattle's "inside" wiremen's local, violated Title VII of the 1964 Civil Rights Act. A detailed account of what led to this ruling and its aftermath , including some of Nicole Grant's work, can be found at the University of Washington's online Seattle Civil Rights and Labor History Project.

Court-mandated changes to #46's apprenticeship program allowed Janet Lewis and Beverly Sims to earn journeymen's cards as the first women in #46 to do so, in 1972. Groundwork for SCL's ETT program was laid in August of that same year, when then-Seattle Mayor Wes Uhlman issued an executive order establishing an affirmative action program for all city departments, including Seattle City Light. Within a year Seattle extended employment protection to lesbians and gays.

In 1973, Uhlman appointed Gordon Vickery Superintendent of Seattle City Light.

> Mayor Wes Uhlman, fearful that former Fire Department Chief Gordon Vickery would run against him, appointed Vickery Superintendent of Seattle City Light in an attempt to placate his rival. Ever the politicians, Vickery and Uhlman decided it would be better to voluntarily hire a few women into non-traditional trades at Seattle City Light rather than risk the type of lawsuits brought against the construction unions and contractors by black workers in the late 1960s. They wanted to do a training program for women and they wanted it to succeed where previous ETT programs targeting low-income and non-white male workers had largely failed in the past. Clara Fraser, who was well known for her feminist activism by this time, was recruited by Vickery to design the new ETT program.
> Nicole Grant, "Challenging Sexism at City Light: The Electrical Trades Trainee Program"

In 1974 Seattle City Light began implementing what would be its first and last Electrical Trades Trainee Program. Nicole Grant (currently executive director of the Certified Electrical Workers of Washington) describes the general social situation leading to this program:

> [T]he ETT program began in an environment charged by racial conflict, as well as worker combativeness and solidarity. The political period of the 1960s and early 1970s wasn't just about the civil rights movement, or feminism or the anti-war movement, either. It was a time of political radicalism on all levels across the nation. Many of the white male workers at Seattle City Light were themselves wrapped up in the era's up-heavals and expressed themselves through job actions whose militancy was much greater than today's.
> Nicole Grant, "Challenging Sexism at City Light: The Electrical Trades Trainee Program"

More than 400 women applied to the ETT program. Ten women were selected. Fraser designed the two-week program to provide theoretical, practical, and physical training. She insisted the program bring all ten women on together, and that they become union members when they began the program. The ETT participants had their own bargaining unit in #77 from the beginning. (John Starcevich was business manager and Art Dakers, Jr. was president of #77 at this time).

At the time, #77 represented some women at SCL, in its metering department for example, but the majority of SCL's office workers were not yet represented by any union.

Local #77's members at SCL staged a ten-day walkout less than two months before the ETT program was to start. Called the "Coffee Break Walk-out" in the press, it was prompted by SCL's disciplinary code. Most of SCL's non-union clerical (and female) workers also walked out in support. Fraser was also active in her support of the striking workers while some of the ETT trainees also voiced their support. All SCL employees came back to work eleven days later, after management agreed to several major concessions. (See the chapter on Starcevich's term as #77 business manager, page 85, for details about this strike.)

Animosity between workers and management was still strong when the ETT women began working for SCL on June 24, 1974. When SCL management cancelled the ETT program after one week, it was clear to all that Vickery was taking action against Fraser and anyone who had supported the walkout.

Fraser was removed as Training Coordinator, and the ETT women were told to go out into the field. Heidi Durham, Kathleen Merrigan, and Megan Cornish, all ETT trainees, remember clearly how quickly Vickery's attitude shifted from his initial desire to be seen as promoting the ETT program.

> Upon getting word that their training had been canceled, most of the trainees, led by Daisy Jones, the oldest and a natural leader, marched up to Vickery's office and demanded an explanation. Dissatisfied with his response, they soon filed a complaint with the city's Office of Women's Rights saying that they were being denied the full two weeks of training that male trainees had received in previous years.
>
> Weeks later [after training], an employee meeting was called by the "coffee break walk-out" leaders to rally for demands that were still unmet, including the ouster of Vickery. Most of the trainees attended, and a few made statements of solidarity with their fellow workers.
>
> Vickery called the women into the office the next week. Stating they didn't seem very happy and were causing a lot of trouble, he proceeded to hand out a loyalty oath for each to sign promising that she would carry out any duty required by management without complaint if she wanted to keep her job. In the end, those who had not already signed added a paragraph to the document stating that by signing they were not relinquishing their constitutional rights, before submitting the signed oath to management.
>
> Nicole Grant, "Challenging Sexism at City Light: The Electrical Trades Trainee Program"

Most of the women went to work as trainees (pre-apprentice workers), not having completed the full ETT course beforehand. The following summer most of these women were laid off, along with Clara Fraser. The women trainees added their dismissals to complaints already on file with Seattle's Office of Women's Rights. Several kept working at SCL in different positions, and some eventually became apprentices and journey-level workers. The ETT program was formally ended in 1976.

It would take until 1982 before Clara Fraser got her job back, and another two years for her to win damages, back pay, and attorney's fees totaling more than $135,000. Only in 1984 did SCL hire another woman as a lineman trainee, and that hiring was the result of another wave of feminist challenges to SCL's hiring practices. Not all of #77's staff supported this later push for affirmative action.

> At the time of their retirements, IBEW Local 77 held a special ceremony for the ETT participants [those who had continued to work as #77 members] thanking them for their groundbreaking work and long records as good union members. The plaques they were presented with were addressed to "Our Very Own Radical Socialist Bitches," referring to an incident when a bigoted [#77] union official was busted calling the women that in the late 1980s. After hearing of their "nickname", they showed up at the next union meeting with boxes full of "Radical Socialist Bitch" buttons and by the time the meeting was brought to order nearly every union member in attendance was sporting one.
>
> Nicole Grant, "Challenging Sexism at City Light: The Electrical Trades Trainee Program"

Left: ETT candidates and Gordon Vickery, superintendent of Seattle City Light, meet the press as the program begins in 1974. (L-R) Jo ann Simmons (standing), Clara Fraser, Marge Belinger (behind Vickery), Gordon Vickery, Letha Neal, (Jody Olivera hidden), Jennifer Gordon. Kathleen Merrigan Collection.

Above right: One of the ETT's, Angel Arrasmith (or Arrowsmith), made journeyman, construction, working out of SCL's Service Center. Her retort to male harassment: "I'm more of a woman than you'll ever get, and more of a man than you'll ever be."

Clockwise from immediate right: Teri Bach and Daisy Jones. Letha Neal. Megan Cornish (facing viewer) teaching at pole climbing school. Daisy Jones. Kathleen Merrigan Collection.

"Our Very Own Radical Socialist Bitches"

This phrase became an inside joke between #77 and the women who remained #77 members after the ETT program. A number of these women were also members of the Radical Women, a feminist group in Seattle. When a bigoted #77 staffer called them "radical socialist bitches" at a union meeting, they returned with boxes of buttons with this slogan and many men within #77 left wearing the buttons.

Teri Bach was the first woman to become a journeyman lineman at SCL and its first woman cable splicer. Her neck was broken in a horrific industrial accident. (Bach died in 2005.)

Daisy Jones quit SCL to take a job at King County Metro driving a bus while the ETT's were off the job because she needed some form of income. As a result, when the women were finally rehired, Jones was no longer eligible for the program because she had technically quit under economic duress.

Letha Neal, the first and only black, female cable splicer at SCL was also the last of the ETT participants to turn out of the apprenticeship and reach journeyman status. (Neil died in 2005.)

Jody Olivera, Chicana, became a substation construction, then went into work safety (out of the trade).

Patti Wong openly criticized the militancy of the ETT program to the local media. She was not laid off with the others, and received part of the settlement from SCL although she was not in the suit. Wong became a cable-splicer helper (non-journeyman).

Megan Cornish became SCL's second woman power station operator.

Jennifer Gordon, like Wong, was not laid off but she quit before the others came back. She quit SCL soon after because she felt discriminated against. Gordon was the top-scoring participant on the civil service proficiency tests. When the ETT's were laid off Marge Wakenight went back to her job as an administrative secretary at SCL.

Right: After her work-related back injury, Heidi Durham fought to come back to work at SCL. She became a power station operator, then the first woman power dispatcher. Kathleen Merrigan Collection.

Below: A partial listing of first women journeymen at Avista. Avista has its own JATC-approved apprenticeship program. Information courtesy of Suzanne Brunner, #77 assistant business manager (Spokane office).

Name	Title	Journeyman Test Date
Suzanne Brunner	Journeyman Hydroelectric and Substation Operator	(1st Journeyman / 2nd Apprentice Journeyman) 12/3/1983
Monica Pierce	Journeyman Communication Technician	1st Apprentice / 2nd Journeyman 1984 (delivered twin girls during Fourth Step of her apprenticeship)
Gayle Taylor	Journeyman Hydroelectric and Substation Operator	Journeyman Test 10/15/1985
Nancy Trevison	Journeyman Hydroelectric and Substation Operator	Journeyman Test 9/30/1990
Kimberly Mattern	Journeyman Hydroelectric and Substation Operator	Journeyman Test 9/30/1990
Judith Moore Allenfort Robertson	Journeyman Mechanical / Structural	Journeyman Test 11/13/1997
Eddie Sue Judy	Journeyman Hydroelectric and Substation Operator	Journeyman Test 9/14/2006
Joyce M Cardwell	Journeyman Gas Serviceman	Journeyman Test 1/14/1999
Michelle Buehler	Journeyman Meterman	Journeyman Test 5/21/2004
Kim Varner	Journeyman Gas Meterman	Journeyman Test 1/18/1992

Customer Service Representatives: The Face of the Utility

Local #77 has a tradition of representing both craft (apprenticeship-trained) and non-craft utility workers. This distinction by job classification is reflected in the IBEW's A and B membership categories, which the entire IBEW first established in 1946. "A" members are generally construction linemen. "B" members pay full per capitas (individual union dues) and receive full voting rights with the exception of any pension- or death benefit-related issues, nor can they vote on these matters (B members have at some times also been identified as BA members).

Historically, the rate of pay for #77's non-craft members at any particular utility was negotiated as a percentage of linemen's wages at that utility. This was both a function and indicator of workplace solidarity across the membership. Over time, distinctions between the day-to-day duties of #77's linemen and non-craft members have become more significant. These changes have contributed to some cases where the wages of non-craft members are no longer pegged to those of linemen.

While linemen still work in crews and have a strong sense of personal connection with their immediate coworkers, many non-craft members are spread across workplaces with a variety of job classifications. In addition, linemen and non-linemen may not have frequent contact with one another even when they share an employer. All of #77 members' jobs are being restructured as utilities, even publicly owned utilities such as PUDS, push for ever increasing profits. Because of the rigors of their craft training, many journeymen retain job security even today, while non-craft workers face ever greater automation, and in some cases, are seeing their work contracted out.

This article will describe the working conditions of one group of #77's non-craft members, the customer service representatives (CSRs) at one utility, Snohomish PUD (SnoPUD). Recent changes in CSR working conditions at SnoPUD illustrate many of the issues and history particular to non-craft members of #77.

First of all, some clarification on terminology relating to SnoPUD's CSRs. At SnoPUD, if you walk into the SnoPUD offices to pay your bill or discuss your account, you will meet with one of some twenty-five to thirty CSRs. Additional CSRs work nearby but not at the counter, on accounting and other specialized tasks such as assessing or waiving deposits, connecting customers with charities that help with power bills, processing accounts that have been included in bankruptcies, and working with builders or owners of multi-unit complexes on metering issues.

SnoPUD's Customer Account Representatives (CARs) work in a separate location from the CSRs. CARs, considered a different classification within the same bargaining unit as the CSRs, receive the same rate of pay as CSRs but perform other tasks (including billing), and currently do <u>not</u> take calls from customers.

Other CSRs work away from the front office. Some thirty to forty CSRs work in the main SnoPUD call center at any given time. Known as "the cellar," this call center is located in a large open basement area dominated by a huge digital wall display. (The center and display have been temporarily relocated due to recent flooding.) The digital display constantly shows the number of callers on hold, the length of time waited for the call that has been holding the longest, the number of CSRs currently on calls, the number of CSRs "logged in" but not currently on calls, the number of calls "offered" (i.e., the total number of calls coming into the queue on a given day regardless of how, when, or if they were answered), and the total number of calls abandoned (callers who entered the queue but hung up before a CSR responded). The same information is also displayed on every CSR's computer screen, no matter what their work duties. There is no monitoring of the complexity of a call's request, nor of the customer's satisfaction.

At other utilities, the various duties outlined above are performed by workers with a variety of job titles. At many Rural Electric Associations (REAs) these workers are called member service representatives; at some PUDs they may be called cashiers or consumer service representatives.

Above: Telephone operators. IBEW IO archives.
Right: Customer service operators at PSP&L.
University of Washington Libraries, Special Collections,
UW35551.
Below: Telephone operators, circa 1930s. IBEW IO
archives.

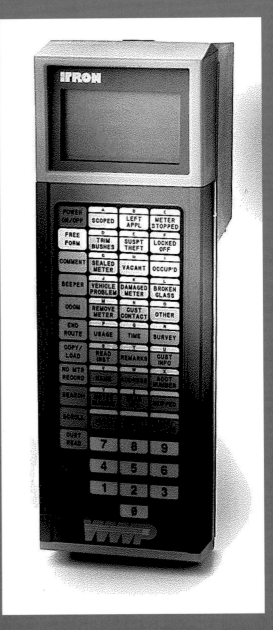

A page from a 1918 WWP meter-reading book, recording electrical use by a business customer with 16 outlets who used 10 kilowatts from January 21 to February 20, 1918. Avista Archive.

```
CW   LCW   ACD  NR  AVL    ANS/ABN
 2   0:16   11   8   0      909/7
```

The digital display used by CSRs at SnoPUD.
  CW = calls waiting.
  LCW = length calls waiting, the amount of time (in minutes and seconds) earliest caller in queue has been waiting.
  ACD = active call detail, the number of CSR's who are on active calls with customers.
  NR (in yellow) = not ready, the number of CSRs who are doing after-call work.
  AVL = available, the number of CSR's sitting in the queue actively waiting for a call to come in.
  ANS/ABN = answered/abandoned, the number of calls answered by CSR's/number of calls that hung up before a CSR could answer. Photo courtesy of Nichole Reedy.

Above: A hand-held Itron meter reader used by WWP, one of the first to use telemetry. Local #77 Archives.

Below (L-R): Karissa Byrne and Lola Wells at the 2012 Customer Service and Call Center Summit. Local #77 Archives.

There are several other job classifications whose work may be related to that of CSRs, or workers with job titles that are sometimes confused with CSRs.

SnoPUD employs energy conservation consultants, who recently organized to gain #77 representation. They conduct technical field inspections relating to energy conservation, and earn substantially more than CSRs. They are also predominantly male, while CSRs are predominantly female. There are also approximately thirty office support specialists (OSS) strewn across the company's officers, explains Nichole Reedy, #77 business rep. "Some are cashiers (taking payments, but cannot discuss accounts), some answer the console ('Thank you for calling Snohomish PUD, how may I direct your call?'), and others perform background work including updating forms, sending letters, filing, and more."

Dispatchers direct, or dispatch, line crews for fieldwork, and may direct fieldwork using computerized energy system mapping. As utilities upgrade their technologies, dispatchers may have telemetric control over energy transmission infrastructures. Historically, dispatch was performed by the workers who answered customer service calls regarding outages, but now dispatch is considered a separate duty, and also a craft.

Energy control dispatchers, or system operators, do the real-time power trading of their utilities' generating sources and transmission systems as part of the larger regional power grid.

The metering department was originally separate from the CSRs, and included meter readers. Meter readers traditionally maintained and repaired meters, and also visited each service meter on a monthly basis to record usage. In some cases they delivered bills, accepted payments, or turned service on and off at the meter.

As telemetry and other metering technology advanced, meter-reading has become less and less about site visits or even neighborhood visits, and more about processing collected data. In the case of SnoPUD, metering functions other than bench maintenance are now automated. Site visits to a particular residential or commercial user are limited to shutting off (or turning on) service at the meter, and are now handled by customer service field representatives, CSRs who are working in the field area of customer service.

At Puget Sound Energy (PSE), meter reading is contracted out, and is not represented by #77. Avista's CSRs and many of its other non-craft members do not have union representation, and have not for many years. Local #77 has waged several organizing campaigns at Avista in recent years without securing new units.

Modernization has driven significant changes in CSR work. Again, using SnoPUD as an example, numerous advancements in call center technology have increased the stress of CSR work. The constant display of backlogged customers, along with a policy of minimizing the length of any single customer call, creates the worksite pressure typical of call centers. Personal and lunch breaks are staggered so that individual workers are isolated from their coworkers. "I feel chained to my desk whenever I am here," says one CSR. Work shifts in SnoPUD's CSR cellar are chronically understaffed, making it extremely difficult for workers to use the vacation time they have earned. At SnoPUD, a CSR can cash out such unusable vacation when it exceeds sixty vacation days, the equivalent of four years worth of vacation. PSE has similar staffing issues regarding vacation, but maintains a use-it-or-lose-it policy.

In August 2012, SnoPUD changed the job description of field CSRs in ways that added to the physical danger of their work. Now, when shutting off a customer's electrical service, field CSRs do not go to the door first to alert the customer that they are there, or why they are there. They no longer carry any account information with them, making it extremely difficult to answer customer questions onsite. In addition, field CSRs can no longer accept any onsite payment. (Local #77 has been lobbying in Olympia to increase the charge for assault of any utility worker from assault to assault in the third degree.)

These changes have made SnoPUD's CSR fieldwork disagreeable enough that field CSR positions are being filled by reverse seniority. Field CSR work is still voluntary at SnoPUD, so some CSRs are "declining" formal field training, preferring to receive less pay rather than go out and perform such non-communicative shutoffs.

In response to the worksite isolation and other job difficulties faced by CSRs, #77 began holding annual CSR Summits in 2007. (They are now called Customer Service and Call Center Summits.) Then-business manager Don Guillot and business reps John Cunningham and C.O. Smith hatched the idea. From the beginning, the goal was a unionwide conference that provided time for CSR workers, especially shop stewards, to discuss and compare their working conditions.

At the first CSR Summit, delegates consisted of whichever CSRs were able to attend for a full day on Friday and a half-day on Saturday. Attendance has increased each year, with the percentage of shop stewards in attendance also increasing. Topics of the CSR Summits have included attorney presentations on the Family and Medical Leave Act (FMLA), Labor and Industries (L&I) claims, social media and worksite computers, workstations that require workers to stand, and potential employer health incentives for desk-bound CSRs. The Customer Service and Call Center Summits have become ever more important as these #77 members face increasing automation and worksite isolation.

Top: Milt Strong working on an ice-covered Inland Power pole on Mt. Spokane in the 1940s. Rick Luiten Collection.
Above, left and right: Snow-packed powerlines, fire lookout and KXLY TV station (lump at base of lookout) on Mt.
Spokane, 1940s. Rick Luiten Collection.

Recent Accomplishments

THE WASHINGTON WATER POWER COMPANY

Electric and Natural Gas Service

P.O. BOX 3727 • SPOKANE, WASHINGTON 99220 • (509) 489-0500

June 3, 1974

Union Negotiating Committee
I.B.E.W.
North 1506 Washington
Spokane, Washington 99201

Union and Company negotiators met 16 times between February 5 and April 10 of this year and have met subsequently an additional eight times under the auspices of a federal mediator. Wednesday will mark the beginning of the sixth week of a strike which your Union called against The Washington Water Power Company. We are sure you agree that we must all take every reasonable step to resolve our differences and get all employees back to work at the earliest possible date.

Obviously it is difficult for both you and the Company to accurately assess the attitudes and desires of Union employees who are on strike. Many Union employees have indicated to us a desire to return to work. They should not have to be reminded that the Union called the strike against their company and that the company has not closed its doors against any employee. Further, we have not hired any permanent replacement for any striking employee. Numerous employees have contended that the strike vote was secured only to strengthen the Union bargaining position. Others say they were assured that a strike, if one were called, would not last more than one or two days. Some have expressed resentment at not having been given the opportunity to vote on Company offers. Of course, some employees stand fast behind your demands. The Union belongs to the employee membership, and we cannot properly interfere in nor suggest how it conduct its business.

Despite the Union demand for $.75 on the 100 percent rate from which you have not deviated, the despite the fact your Union chose to call a strike on this issue, the Company has continued to bargain in good faith and made several different offers to you which you have not chosen to pass on to the membership for vote. It is for the above reasons we, today, urge you, the Union Committee, to take the Company offer of June 3, 1974, to your membership so that they can reflect their views by secret written ballot.

This letter is written to you as the Committee which represents our Union employees. A copy of this letter, together with the Company offer, is being sent to all employees only for their information. By informing all interested employees of this matter, we feel that they can more intelligently provide you with instructions as to their collective attitudes and desires.

The Company Negotiating Committee

By: William R. Yeo
Secretary

Building a Powerful Future

THERE is a point in writing history where events are so close to the present it is impossible to provide historical perspective. The following is a compilation of significant achievements at #77 since 2004. All of these accomplishments make it clear that the membership, executive board, business representatives, and staff of #77 recognize the need to work for the common good of the local's members. Together, their talents and their commitment to #77 and the union movement ensure the local's continued success.

Financial Stability

Local #77's financial situation was precarious when Don Guillot was elected as Business Manager in 2004. At that time, #77 was putting little into savings. The local's savings plan has improved considerably since then. (A significant dues increase at the end of Dave Timothy's term helped make this possible.)

Local #77 purchased and operates the Northwest Utility Training and Education Center (NUTEC) and made the move to its new office building in 2013 without a dues increase, as Guillot promised the membership.

Improved Membership Representation

The number of business representatives at #77 has steadily increased since 2004.

The local now uses a mentoring program to train new union leaders. Rank-and-file members work under the direction of a business representative.

While attending shop steward training and training in negotiations, members' wages are now covered by #77. This allows more members to receive such training.

Legislation and Political Action

(Additional information about #77's actions in the state legislature may be found in "Defending Workers in the Legislature," page 144.)

In the mid-2000s, #77 began to be more politically active, and more politically effective, than it had been in years. In the decade since, the #77 PAC has become a significant and effective tool for political action, interviewing political candidates to determine endorsements, and making key campaign contributions. Local #77 has successfully lobbied for the following recent legislation:

    In 2005

        HB 1557 – "Lineman on board"

    In 2007

        HB 2391 – PERS "incremental improvement"

        HB 2639 – "Standard wages to be paid on all renewable generation projects built by PUDs"

        HB 2171 – "Crane workers must be qualified and certified"

Left: The letter George Bockman received from WWP during the WWP strike in 1974. Bockman is a retired member of #77 who served on #77's strike committee. George Bockman Archive.

Programs

IBEW Local #77 Sponsorship Program

Started in 2004, this program provides funding to programs and events in which 377 members and their families participate. The program promotes recognition for #77 by displaying #77's logo and/or name on program materials and/or uniforms.

Customer Service & Call Center Summit (CSR Summit)*

In May 2007, #77 presented its first conference for customer service representatives (CSRs). The CSR Summit is now an annual event, providing perhaps the only opportunity for CSRs and call center workers to compare notes and discuss workplace issues. Given the many rapid changes in related technologies and job classifications, the CSR Summit demonstrates #77's commitment to fostering workplace leadership for all its members.

NUTEC

(Additional information about NUTEC may be found in the chapter, "Defending Workers in the Legislature," page 144.)

The Northwest Utility Training Education Center (NUTEC) is the only training facility focusing on journeyman level, hands-on training within the energy industry. In November 2009, #77 purchased the entire seventy-seven-acre site near Richland and immediately conveyed approximately five acres to Hazardous Materials Management and Emergency Response (HAMMER), a federal Department of Energy program. (Guillot thought #77 buying seventy-seven acres was an auspicious sign.) Training at NUTEC is designed by journeymen to address what they identify as immediate training needs. In 2012, military veterans participated as observers in #77's pilot training program.

A New Business Office

By December 2009, #77 had outgrown its existing building and began to look for a new office. Local #77 purchased its new hall in 2012 without a dues increase to members. Local #77 moved to its new SeaTac location at 19415 International Boulevard South in April 2013.

Local #77's new building (19415 International Boulevard S., SeaTac) in spring 2013. Photo by Kellie McGuire.

Right: The Andrew York Linemen's Rodeo honors Chelan County PUD journeyman lineman Andrew York. York died in 2000 from injuries suffered on the job when he was hit by a drunken driver. The rodeo is hosted by the Chelan, Douglas, Ferry, Grant and Okanogan PUDs and IBEW #77. Local #77 Archive.

In 2006, #77 sponsored "The Transformers," a Marysville Little League team. L-R front: Jack Allen, Austin Pygott, Ethan Radford-Chen, Kyle Trapp. L-R rear: Jesse Belmont (coach), Alex Belmont, Cece Watson, Alex Rodriguez, Sativa Mock, Abbie Allen, Jeff Allen. Local #77 Archive.

Safety

The IBEW #77 Safety Committee initiated field visits to every construction site within 77's jurisdiction in October 2007.

Local #77 now runs an accident investigation program in which #77 volunteers go onsite to investigate accidents. The program emphasizes gathering accurate information rather than assigning blame. In no way do #77's investigations alter or replace Washington State and/or employer fact-finding or other practices related to accident investigation.

Local #77 participates in two safety groups:
The Electrical Utility Safety Advisory Committee (EUSAC), a joint effort by the industry and IBEW to review and make recommendations regarding state and federal industrial safety rules.

TASKFORCE TEAM 3, a recent collaborative effort within the utility construction industry comprising representatives from the IBEW, the National Electrical Contractors Association (NECA), union and non-union contractors, the Occupational Health and Safety Administration (OSHA), the Washington Department of Labor and Industries, and the Joint Apprenticeship Training Committee (JATC). TEAM 3 works to reach agreement on best practices in construction safety for the entire construction industry. Once agreement is reached, this single set of concepts will be enforceable through OSHA.

NECA-Four Local Agreement

("NECA-Four Local Agreement" refers to the agreement between the Northwest Line Constructors Chapter of the National Electrical Contractors Association (NECA) and four regional IBEW linemen's or construction locals, #77, #125, #483, and #659.)

The early 2007 NECA-Four Local Agreement contract secured "one of the largest wage increases and benefit improvements in the last twenty years," according to the report that #77 submitted to *The Electrical Worker* in May 2007.

In 2009 the NECA-Four Local Agreement initiated Health Reimbursement Accounts (HRA) for represented workers.

The 2013 NECA-Four Local Agreement includes short- and long-term disability insurance that will become effective in 2013.

The 2010 "Pink Pistols," a roller derby team, were sponsored by #77. Local #77 Archive.

Defending Workers in the Legislature

Local #77 has played a significant role in shaping Washington State's laws. Below is a list of some of the local's recent legislative victories, some potentially harmful legislation that #77 helped to defeat, and future legislation that is a priority for #77.

Bob Guenther, a staunch supporter of labor, works to make sure #77 is represented in Olympia. As #77's governmental affairs advisor, Guenther serves on the legislative team of the United Labor Lobby (ULL). This coalition, which includes the Washington State Labor Council (WSLC) and other community partners, supports legislation that improves the lives and job conditions of working people.

Below are highlights of recently passed legislation.

2005
House Bill 1557
2005 Governor's Electrical Board
Expands membership of the electrical board by appointment of one outside lineman. Former #77 Business Manager Don Guillot was the first to serve in this position, and served two terms.

2006
House Bill 2538
Industrial Safety and Health Act—Inspections
Authorizes Washington State Department of Labor and Industries (L&I) to seek warrants for Washington Industrial Safety and Health Act (WISHA) inspections at job sites, allowing L&I access to the worksite.

Senate Bill 6885
Restores the averaging calculations that determine the amount of an unemployment benefit back to using two quarters of employment rather than figuring unemployment rates based on a four-quarter average. The restored calculation generally gives the unemployed person a larger unemployment check.

2007
Senate Bill 6014
Modified the Growth Management Act (GMA) to allow industrial development on the Centralia Mine site.

House Bill 2391
The retirement age of PUD employees was lowered to 62 without penalty. *This bill was revised in 2013 so that now only those vested under the 2006 legislation have this option; new hires do not.*

2008
House Bill 2639
Requires that renewable energy projects by PUD and other entities comply with state prevailing wage.

Senate Bill 6560
Increases bid limits for PUD employers bidding on PUD work, from $50,000 to $150,000. This means PUDs can bid on larger projects. Legislation was initiated by #77.

2009

Senate Bill 5492

Adds nuclear power plant workers to the employees covered by interest arbitration under the Public Employees' Collective Bargaining Act. Specifies factors to be considered by an interest arbitration panel resolving an impasse in collective bargaining involving these employees. If management or the union declares impasse, negotiations now go into arbitration.

2010

House Bill 1323

Codifies the Community College Center of Excellence Concept for Cluster Industries. This standardizes curriculum so various community colleges within the Center of Excellence program can offer the same courses.

2011

House Bill 1618

PUD Deferred Compensation: All elected PUD commissioners receive a wage increase.

Senate Bill 5769

Keeps TransAlta's Centralia Generation complex operating until 2025, providing significant economic growth, including union jobs, in Lewis County.

Current Legislative Issues

Local #77 is working to increase the charges for assault of any utility worker from "assault" to "assault in the third degree." This change addresses the increasing risks that utility workers face in interacting with the public, especially when they cut off service onsite.
(See "Customer Service Representatives," page 134, for additional information.)

Local #77 is working for new legislation that would recognize linemen as "first responders" under Public Employees' Retirement System (PERS).

Recent Legislation Blocked

Local #77 worked to block efforts that would require state licenses for linemen. Although inside electrical workers (wiremen) are licensed by the state of Washington, journeymen linemen have mutual aid agreements in the face of natural disasters that allow them to work throughout the United States. Individual state-licensing requirements would severely impede the linemen's ability to respond at levels sufficient to address such emergencies.

NUTEC Leverages Expertise

Local #77 currently owns and operates the Northwest Utility Training and Education Center (NUTEC), a training facility outside Richland, Washington, which allows #77 to offer its jounrneymen linemen supplementary training. NUTEC began under the auspices of the Northwest Public Power Association (NWPPA), a trade association of PUDs, electric cooperatives, municipalities, and crown corporations in the western U.S. and Canada. Through partnerships with Bonneville Power Administration (BPA) and some 200 public utilities, NWPPA established its National Utility Training Services (NUTS) training center where NUTEC now operates. NWPPA acquired the site in 2003 from the U.S. Department of Energy (DOE) to be developed as a state-of-the-art training facility for line, substation, meter, and relay personnel, along with electricians, engineers, and office personnel.

At some point NUTS was renamed NUTEC. When NWPPA leadership changed, it cut funding for the facility and "everything went defunct," according to #77 business rep John Trumble. NWPPA's reasons for shutting it down are not clear. The facility sat vacant for several years while a number of individuals and organizations looked for a way to put the training center back in service. After several poorly funded collaborative attempts which included #77, the site was returned to the DOE's Department of Education.

By 2004, when Don Guillot was business manager, #77 was interested in acquiring the NUTEC facility for its own training facility. The local began discussions with IBEW's International Offices (IO). But #77 and the IO had different concepts for training facilities: the IO's model was to lease a property, then work with a community college. Local #77 wanted to own the property outright and take charge of programming and training. Local #77 had the money.

With the approval of the IO, the local bought the entire facility in November 2009, keeping the name NUTEC. (Confusion between #77's NUTEC and the failed original NUTEC has created some challenges for #77.) Local #77 immediately transferred ownership of approximately five acres back to the DOE. This property became part of the DOE's Hazardous Materials Management and Emergency Response (HAMMER) center.

Bob Topping, an educator in the construction industry, had not previously designed linemen's training. Topping began developing a hands-on training process whose content would be designed by journeymen linemen. Based on talks between #77 members and Topping, NUTEC programming came to focus on "young leaders," 35- to 50-year-old linemen who are also worksite leaders. These linemen field-train new apprentices coming onto the worksite. Apprentices may know more than the journeymen do about emerging technologies, but as apprentices, they do not have the field experience or employer-specific knowledge of their seniors.

Arlene Abbott was hired "to open the doors" and succeeded in securing a series of small grants. The first, during the summer of 2011, paid for "young leaders" from within the Columbia River region who met and identified specific hands-on training that would improve their own work. Their recommendations were then rolled into a prototype program.

The first NUTEC training session, in January 2012, covered safety and hazard analysis. A more complete pilot program, Underground Systems, focused on troubleshooting energized systems, and used journeymen as educators and trainees. This program ran from June 2012 through January 2013. Its funding also allowed NUTEC to reach out to military veterans in Washington's Benton-Franklin PUD area. Vets spent two days per training session observing what linemen's work entails.

Rick Irvine, NUTEC instructor, gave the orientation at one of the pilot program sessions and explained that NUTEC is designed by linemen, for linemen, to develop "best practices."

> Best practices really come down to a specific situation. They're based on the way work is done, what equipment is available, what the relationship is to dispatch, whether it's maintenance work or an emergency.
> Rick Irvine, NUTEC training session, 2013

NUTEC's current governing board is the executive board of #77. NUTEC also has an advisory board with representatives from #77, various Central Washington PUDs, the Pacific Northwest Center of Excellence for Clean Energy (PNCECE), the Washington Labor Council, and workforce and economic development organizations.

Above: The NUTEC facility, 2012. Photo courtesy of Candice Bluechel, Director of NUTEC.

Below: The NUTEC logo. Photo courtesy of Candice Bluechel, Director of NUTEC.

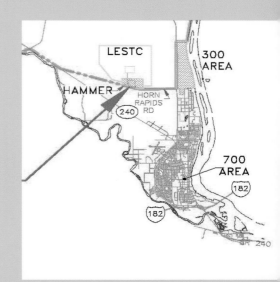

Above: A journeyman lineman works a hot stick as part of a NUTEC exercise in troubleshooting as a military vet observes, spring 2013. E. Belew photograph.

Right: NUTEC is located northwest of the TriCities, immediately adjacent to the HAMMER facility. U.S. Department of Energy.

Richland crew doing hot work, circa 1953. They are working on three-way service, using a hot stick, to rig a temporary cross-arm. Top L-R: John "Buster" Perry, Bob "Big Foot" Smith. Lower: Jim Herndon, who was a third-year apprentice at the time. Buster Perry Collection.

Appendices

Contract Units

THIS table represents a distillation of available information, and with luck will serve as a basis for further research and compilation. Local #77 members are represented in units, or groups, which may span employers and/or locations and/or job classifications.

The more sources that were consulted, the more likely it was that discrepancies or incomplete information would appear. The #77 Collection at the University of Washington Libraries, Special Collections, includes many of the original agreements. Each of #77's offices (Seattle, Kennewick, and Spokane) has files with references to, or copies of, or correspondence relating to various contracts. In addition, both #77's newsletter and *The Electrical Worker* make references to various units and agreements, dating back to the 1940s. Occasionally a photograph contained a unit number. In some cases, an employer's name was only cited by acronym.

KEY

? indicates a portion of the information listed could not be confirmed but was cited in at least one source.

***** indicates a unit that includes more than one employer.

****** Unit 111 has historically represented combined PUDS (Franklin, Benton) and some Eastern Washington construction.

HAMTC? indicates that it is not clear whether this employer was, or is, covered under the Hanford Atomic Metal Trades Council (HAMTC) agreement.

Outside Construction (Construct)

Each of #77's three offices has a separate set of books from which it dispatches work: Seattle, Kennewick (Mid–Columbia), and Spokane.

All employers who use this dispatch system are signatories to the NECA–Four Local Agreement. (NECA–Four Local Agreement refers to the agreement between the Northwest Line Constructors Chapter of the National Electrical Contractors Association (NECA) and four regional IBEW linemen's or construction locals, #77, #125, #483, and #659.)

Potelco was a signatory to this agreement before it became a PSE contractor, and continued under the agreement. (The chapter on David Timothy explains in detail how this came about.)

Tree Work (Construct–NECA–Tree)

There is a wage schedule for tree work in the NECA–Four Local Agreement, but it is not used. Asplundh has its own contract with the four locals of the NECA–Four Local Agreement. All other employers whose work is dispatched by #77 have what is called a "me, too" contract, in which the employer must meet or exceed the terms of the Asplundh agreement.

PBX/Telephone Work

Private Branch Exchange (PBX) refers to employers who have or had an internal communications or telephone system, or switchboard.

Hanford Atomic Metal Trades Council (HAMTC)

At Hanford, the HAMTC master contract covers everything under collective bargaining, and is between employers and the Columbia Basin Trade Council (CBTC). The contract has an Appendix A listing specific wages and working conditions for specific affiliates, including #77.

Employer	Type of Employer	Contract Dates	Unit No.	Local #77 Office
American Utility & Residential Tree Service	construct-NECA-tree	2010	149*	Spokane
Arbor Tree Surgery	construct-NECA-tree	1997–2001		
Areva NP, Inc. (formerly Framatome) see Siemens Nuclear	HAMTC?	current	141*	Kennewick
Asplundh	construct-NECA-tree	1985–current	149*	All
Asplundh Mowing	construct-NECA-tree	1985–current	149*	All
Associated Arborists	construct-NECA-tree	1996–current		All
Avista–WWP (Clearwater)	utility co	1993–current	126*	Spokane
Avista–WWP (Colville)	utility co	1993–current	137*	Spokane
Avista–WWP (Grangeville–Orofino)	utility co	1993–current	130	Spokane
Avista–WWP (Kootenai)	utility co	1993–current	109*	Spokane
Avista–WWP (Moscow –Pullman)	utility co	1993–current	125	Spokane
Avista–WWP (Spokane)	utility co	1940–current	137*	Spokane
Battelle NW, see HAMTC Pacific NW National Lab at Richland				
BC Electric/Local 213–230		1951–?		
Bellingham Gas		1946–1947		Seattle
Benton County PUD	PUD	1949–current	111**	Kennewick
Benton County PUD, Prosser	PUD	1953–current		Kennewick
Benton County, Franklin County, City of Richland Construction		1993–current	111**	Kennewick
Benton REA	REA-private utility	1949–current	131	Kennewick
Bestway Tree	construct-NECA-tree	1999–2010		Seattle
Big Bend Electric Coop	REA-private utility	1948–2010		Kennewick
Big Bend REA	REA-private utility	2000–current		Kennewick

Employer	Type of Employer	Contract Dates	Unit No.	Local #77 Office
Big Bend REA/Ritzville	REA-private utility	1953–current		Kennewick
Blume (?) System	construct-NECA-tree	1985–1989		
Boyd's Tree Service	construct-NECA-tree	2012–?		
Bremerton Highway Flagging A & BA	construct-flagging	1986–?	133	Seattle
Broadcasting: all radio and TV	broadcast	1953–?		
Chelan County PUD #1	PUD	1949–current	114*	Kennewick
Cheney Tel. Co.		1953–?	123*	Kennewick
Citizens Utility Co.	Idaho	1964–?		
City of Bellevue	municipal utility	1982–current	116*	Seattle
City of Blaine	municipal utility	1995–?		Seattle
City of Cashmere		2007–current		Kennewick
City of Centralia	municipal utility	1973–current	133*	Seattle
City of Centralia, Centralia City Light	municipal utility	1953–current		Seattle
City of Centralia, Centralia City Light engineers	municipal utility	1951–?		
City of Centralia, Centralia City Light, Centralia Dept. of Lighting	municipal utility	1977–1981	133*	Seattle
City of Centralia, Centralia Water/Wastewater	municipal utility	1989–current		Seattle
City of Centralia, Waste water	municipal utility	1951–1998		Seattle
City of Cheney	municipal utility	1995–current	123*	Seattle
City of Chewelah	municipal utility	2005–current		Seattle
City of McCleary	municipal utility	1989–current	105*	Seattle
City of McCleary (combined with Grays Harbor PUD) see Grays Harbor PUD	municipal utility	1997–current	105*	Seattle
City of McCleary Police Officers		?–current		Seattle
City of Richland	municipal utility	1959–current		Kennewick
City of Richland Electric Dept.	municipal utility	1971–1973		

Employer	Type of Employer	Contract Dates	Unit No.	Local #77 Office
City of Sumas	municipal utility	2008–2011		
Clear Water Power		1960–current	126*	Seattle
Clearwater Power & Light		2011–current		
Cochran	construct-NECA	1995–2001		
Columbia County REA, Dayton	REA-private utility	1953–?		Kennewick
Columbia REA	REA-private utility	1951–?		Kennewick
Construction* in and around Seattle	construct-NECA	1953–current		Seattle
Construction* Mid–Columbia (Kennewick)	construct-NECA	1993–current	111**	Kennewick
Construction* north of Seattle, west of the Cascades	construct-NECA	1953–?	155, 142a, 149	Seattle
Construction* Seattle	construct-NECA	1993–?	142	Seattle
Construction* Spokane	construct-NECA	1993–current	129	Spokane
Cowlitz County PUD #1	PUD	1948 (or 1953?)–current	106	Seattle
Cowlitz County PUD #1	PUD	1993–current	106A	Seattle
Cowlitz County PUD #2	PUD	1993–current	106B	Seattle
Cowlitz Falls Hydro Project		2003–current		Seattle
Dall Jack Inc.	construct-NECA?	2000–?		
Dallas Utility VC (?)	construct-NECA?	1995–1999		
Davy Tree	construct-NECA-tree	1985–2001		
Douglas County PUD	PUD	1948–current	114*	Kennewick
East Side Construction	construct-NECA	1991–current		Kennewick Spokane
Edgewater Inn	telephone	1984–1993		
Ellensburg City Light	municipal utility	1948–current		Kennewick
Energy NW Travelers B.U.	HAMTC?	2001–current		Kennewick
Everson Econovac		2003–?		
Farmers Mutual Tel. Co. of Lynden	telephone	1953–?		

Employer	Type of Employer	Contract Dates	Unit No.	Local #77 Office
Ferry County PUD	PUD	1948–current		Kennewick
Ferry County PUD, Republic	PUD	1953–current		Kennewick
Fluor Daniel Hanford Ind. (formerly Westinghouse)	was under HAMTC but "went away"	1998–2010?		Kennewick
Frame Tome		2002–?		Kennewick
Franklin County PUD, Pasco	PUD	1949–1950	111**	Kennewick
Franklin County PUD, Pasco	PUD	1953–current	111**	Kennewick
General Tel. Co.	telephone	1953–1963		
Grant County PUD #2	PUD	1949 (or 1953?)–current	120	Kennewick
Grays Harbor County PUD	PUD	1945–current		Seattle
Grays Harbor County PUD – Engineers (and Cust. Service)	PUD-CSA+ Engineer	2000–current		Seattle
Grays Harbor County PUD #1	PUD	1958–?	105*	Seattle
Grays Harbor County PUD/City of McCleary see City of McCleary	PUD	1993–current	105*	Seattle
HAMTC contract with various Hanford site (specific employers listed)	HAMTC	1949–current	117 and 140	Kennewick
HAMTC–Battelle Hanford	HAMTC?	2003–current		Kennewick
HAMTC–Bechtel Hanford, Inc.	HAMTC	?		Kennewick
HAMTC–CH2M Hill Hanford Group	HAMTC	2003–current	117 and 140	Kennewick
HAMTC–Eberline Serv. Hanford	HAMTC	2002–current	140	Kennewick
HAMTC–Energy NW	HAMTC	1999–current	140	Kennewick
HAMTC–Johnson Controls	HAMTC	2003–current	117	Kennewick
HAMTC–Pac NW National Lab at Richland) (PNNL, formerly Battelle)	HAMTC	?		Kennewick
Hawkes Electric		1991–current		All
Hawkeye		1998–2001		All

Employer	Type of Employer	Contract Dates	Unit No.	Local #77 Office
Heath		2009–2011		
Henkels & McCoy	construct-NECA	1985–current		All
Highway Flagging System	construct-NECA flagging	1980–1989		
IBEX	construct-NECA	2005–current		All
Idaho Co. Water Power Co, Coeur d'Alene/(Grangeville?)	Idaho	1953–current		Spokane
Inland Empire REA	REA-private utility	1945–?		Spokane
Inland Power & Light	REA-private utility	1995–current		Spokane
Inland Vegetation Management, Inc.	construct-NECA-tree	2011–?		All
Interstate Tel Co	telephone	1951–1952		
JTS	construct-NECA-tree	2001–?		All
Kemp West	construct-NECA	?–current		All
Kennewick A		1983–?	141 and/ or 124?	Kennewick
Kennewick A & BA		1980s–?	111**	Kennewick
King County		1998–current	100*	Seattle
King County DOT signal/ communications		2011–current		Seattle
King County Metro/traction rail power		2011–current	138*	Seattle
King County Traffic Signal		1989–current		Seattle
King County/Metro		1989–current	100* ?	Seattle
Kittitas County PUD	PUD	1998–current	108*	Kennewick?
Kittitas County PUD and REA	PUD/REA	1953–current		Kennewick
Kootenai Elect. Coop	Idaho	1995–current	109*	Spokane
Kootenai REA, Couer d'Alene	Idaho REA	1953–?	109*	Spokane
KVA Electric	construct-NECA?	2011–current		All
Lewis County PUD #1	PUD	1953–current	104	Seattle

Employer	Type of Employer	Contract Dates	Unit No.	Local #77 Office
Lincoln		1963–1964		
M&L Construction	construct-NECA	?		All
Mason County PUD #1, #3 (Shelton, Hoodsport), Simpson Timber Co., Shelton	PUD and unknown	1953–current	128*	Seattle
Mason County PUD #3 (currently represented)	PUD	1973–1975?	128*	Seattle
METRO		1993–current	138*	Seattle
METRO (power/traction rail?)		2012–current	138*	Seattle
Modern Electric		1993–current		Spokane
Modern Electric Co., Opportunity		1953–current		Spokane
Mountain States Power CO, Sandpoint	Idaho construct	1953–?		Spokane (ID)?
Neal's Flagging	construct-NECA flagging	?–current		All
Nespelem Valley Electric		?		Kennewick
Northern Lights	Idaho REA-private	1995–?	136*	Spokane
Northern Lights/WWP (Sandpoint)	Idaho REA-private	1993–?	136*	Spokane (ID)
Northern Lights Group A	Idaho REA-private	2000–?	136*	Spokane
Northern Lights Group B	Idaho REA-private	2000–?	136*	Spokane
NW Line Construction NECA see construction	construct-NECA	1946–current		All
Okanagan County PUD	PUD	1953–current	127	Kennewick
Okanagan County Electric Cooperative, Inc.	construct-NECA or REA?	2010–current		Kennewick
Orcas Power and Light Cooperative (OPALCO)		1993–current	134	Seattle
Orcas Power & Light		1989		
Osmose, Inc.	construct-NECA	2001–current		All
Pacific County PUD #2	PUD	1973–?	110*	Seattle
Pacific County PUD (#1?)	PUD	1953–current	110*	Seattle

Employer	Type of Employer	Contract Dates	Unit No.	Local #77 Office
Pacific County PUD/Wahkiakum PUD (currently both Pacific PUDs and Wahkiakum?)	PUD	1993–current	110*	Seattle
Pacific West Utility Service, LLC.	construct-NECA	2009–?		Seattle
PBX	PBX	1953–1965		
Pend Oreille County PUD	PUD	1953–current	148	Kennewick
Peoples' Tel. Co. of Stanwood	telephone	1953–?		
Potelco	construct-NECA-Four Local Agreement			Seattle
Potelco Admin		2012–current	135	Seattle
Potelco fleet op. mechanics		2011–?		All
Powerliners, Inc.	construct-NECA	2008–current		All
PSE call centers (CSRs)	utility co	2011–current	116*	Seattle
PSE–was then PSP&L	utility co	1993–current	112*	Seattle or Kennewick?
PSE–was then PSP&L	utility co	1993–current	116*	Seattle
PSE–was then PSP&L	utility co	1936–current		Seattle
PSE–was then PSP&L	utility co	1953–current		Seattle
PSE–PSPL	utility co	1994–current		Seattle
PSE–PSPL	utility co	1995–current		Seattle
PSE–PSPL (Bellingham/Alger)	utility co	1993–current	115	Seattle
PSE–PSPL (Bothell)	utility co	1993–current	151	Seattle
PSE–PSPL (Bremerton)	utility co	1993–current	119	Seattle
PSE–PSPL (Goldendale)	utility co	1993–current	150	Seattle
PSE–PSPL (Oak Harbor)	utility co	1993–current	146	Seattle

Employer	Type of Employer	Contract Dates	Unit No.	Local #77 Office
PSE–PSPL (Puyallup)	utility co	1993–current	122*	Seattle
PSE–PSPL (Renton and/or City of Bellevue)	utility co	1993–current	116*	Seattle
PSE–PSPL City of Ellensburg/Kittitas	utility co	1993–current	108*	Kennewick or Seattle?
PSE–PSPL/City of Bellevue	utility co	1995–current	116*	Seattle
PSE+ Potelco Admin+ construction/Olympia		?–current	112*	Seattle
Renton A & BA	municipal utility?	1980s–?	116*	
SCL	municipal utility	1989–current	100*	Seattle
SCL (formerly Seattle Dept. of Lighting?)	municipal utility	1953–1994	102*	Seattle
SCL CMEO'S	municipal utility	?		Seattle
SCL Engineering	municipal utility	1989–?	100*?	Seattle
SCL/King County Seattle Eng.	municipal utility	1993–current	100*?	Seattle
SCL/OPNS Operators	municipal utility	1993–?	102*	Seattle
SCL/Skagit	municipal utility	1993–current	107/107a	Seattle
Sea DOT	municipal utility	2011–?		Seattle
Seattle Dept of Lighting (same as SCL?)	municipal utility	1950–1994?		
Seattle Hotel Association		1951–1980?		
Seattle Steam Company (Corporation?)	utility?	1993–current	100*	Seattle
Seattle Transit System	municipal utility	1944–1956		
Seattle Transportation (Seattle Transit?)	municipal utility	2000–?		
Shelton A & BA	municipal utility	1983–?	133*	Seattle
Shelton, WA Joint Power Operation	municipal utility	1959–?		
Siemens Nuclear, Siemens Power (formerly Areva NP, Inc.?)	HAMTC?	1988–2004	141*	Kennewick
Simpson Timber Co., Shelton		1975–current	128*	Seattle
Skagit Valley Tel. Co of Mt. Vernon	telephone	1953–?		

Employer	Type of Employer	Contract Dates	Unit No.	Local #77 Office
Snohomish County PUD #1	PUD	1952–current	118	Seattle
Spokane		1993–current	123*	Spokane
Stevens County PUD	PUD	1953–?		
Tanner Elec		2011–current		Seattle
The Westin Hotel PBX		1993–current	122*	Seattle
Total Energy Services		2011–?		Seattle
Tree trimmers (treework)	construct-NECA-tree	?–current	149	
Trees, Inc.	construct-NECA	1987–current		All
US & A		2000–2001		Seattle
USBR (US Bureau of Reclamation)	federal/USBR	1984–current		Kennewick
USBR/Columbia Trades	federal/USBR	1981–?		
USBR/Grand Coulee Dam A	federal/USBR	1986–?	124*	Kennewick
USBR/Grand Coulee Dam?	federal/USBR	1953–current	124*	Kennewick
USBR/Grand Coulee PHA	federal/USBR	1984–current		Kennewick
USBR/Hungry Horse	federal/USBR	1999–current		Kennewick
USBR/Upper Columbia Area Office	federal/USBR	2000–current		Kennewick
USBR/Yakima	federal/USBR	1991–current		Kennewick
USBR/Yakima & Hungry Horse Dam Project	federal/USBR	1999–current		Kennewick
Utility Systems & Applications (US & A)		1999–current		
Vera Power & Light		1995–current	123*	Spokane
Wahkiakum County PUD #1	PUD	1953–current	110*	Seattle
Wenatchee/Chelan/Douglas Counties		1993–current	114*	Kennewick
West Coast Tel. Co.	telephone	1952–1965		

Employer	Type of Employer	Contract Dates	Unit No.	Local #77 Office
West Side Construction, Flagging Companies (unclear if this is one or two employers)	construct-NECA	1995–current		
Westin PBX	PBX	2011–?		
Westinghouse (Electricians)		1993–?	117*	Kennewick
Westinghouse (Instr. Techs)	HAMTC or separate contract?	1993–?	140*	Kennewick
Westinghouse Boeing		1993–?		Kennewick
Westinghouse Hanford	HAMTC or separate contract?	1988–?	140*	Kennewick
Westside Construction (sic?)	construct-NECA?	2011–?		Seattle
WHC –Hanford	HAMTC?	1993–?		Kennewick
WPPSS	utility co	1950–1951		
WPPSS see HAMTC Energy Northwest	utility co	1988 (or 1993)–?	132	Kennewick
WPPSS/Adm	utility co	1993–?	103	Kennewick
WPPSS/GPNuc (a.k.a. HGP)	utility co?	1993–?	121	Kennewick

~August, 1937   *The Journal of Electrical Workers and Operators* ·   347

Local Union Organized On State Wide Basis

A UNIQUE organization of a local union is recorded on the Pacific Coast. The union is L. U. No. 77, International Brotherhood of Electrical Workers, serving public utilities including both private and publicly owned utilities in the great state of Washington. The pivotal point of operation is Seattle. The local union extends, however, through the great state of Washington in the western part of the state.

Adopting the slogan, "Go Where Electricity Goes" and adjusting its jurisdiction to suit, this great body of electrical craftsmen has attracted wide attention to itself. The cities served are Seattle, Tacoma, Kelso, Longview, Aberdeen, Hoquiam.

· Moreover, this local union, organized on a state-wide basis, serves the overflow of these plants into rural territory. One of the correspondents of the local union has recently described the remarkable development of this organization in the 1937 Yearbook of the Washington State Federation of Labor:

Local Union No. 77, International Brotherhood of Electrical Workers, is organized on a state-wide basis, since it has to deal with utilities serving a large part of the state of Washington. Operating in a narrow field proved a difficult task, and the field was extended in order to better serve the many hundreds of skilled workmen who are employed by the essential utilities in a fast-growing territory.

Following electricity from its base throughout state of Washington makes L. U. No. 77 unique in its scope and function.

---

tablished a reputation for fair dealing and for open negotiation on the questions involved in wages, hours, and working conditions.

PRINCIPAL CITIES OF GREAT POWER STATE CO-OPERATE

The firms and corporations doing business with Local No. 77, are operating in the several fields of service, and the number of men employed ranges between one and 1,000. The satisfying thing is that the conditions of work and the observance of union regulations are just as pleasant in the small operation as in the larger.

With more than one-half of the total

is safe to say the best service in the country is offered to users in the field of Local No. 77.

Through Business Manager Mulkey, who attends meetings of the membership in every locality at least monthly, the membership of Local No. 77 keeps in close touch with problems as they arise and are able to meet them quickly and without irritation on the part of either men or management. It has been found an excellent plan for the business manager to know personally the officials of the various utilities throughout the state, and thus be able to negotiate with them on a basis of common, friendly understanding.

Not only has it been found essential to negotiate through officials who personally know each other, but it has been found expedient to have unit locals directly affiliated with the main office in Seattle, and with local central bodies as well. A centralized administration has thus been established, with the closest of contacts with local problems, with machinery provided for quick action through negotiation. And, it is known generally, that electrical workers only resort to the strike as a last resort. Locally, the strike has never been necessary.

In the radio field a campaign has been carried on to protect the owners of radios, who of necessity must be illy informed as to the technical operation of their instruments. A war has been waged

Local #77 Officers 1897–1945

Year	Business Manager	President	Financial Secretary	Record Secretary
1897		J.M. Bigler	G.G. Jenkins	
1898		J.J. Maitland	G.G. Jenkins	C.H. Randall
1899		J.J. Maitland	G.G. Jenkins	S. Curceck
1900		John Argutter	D.H. Alexander	George R. Cooley
1901		S.H. Metcalf	George W. Walters	Daniel Sullivan
1902		A. Wagner	George W. Walters	G.W. Davis
1903		James M. Bateman	Jim Brown	Archibald Gordon
1904		John S. Wilson	C.A. Young	C.J. Knago
1905		B.W. Bowen	George W. Walters	Archibald Gordon
1906			W.A. Trousdale	L.O. Andres
1907			W.B. Reed	R.H. Sylvester
1908			W.B. Reed	R.H. Sylvester
1909–1913		George L. Brooks*		
1914		George L. Brooks*	R.W. Hemming, (Treasurer D.J. Lundy)	Frank Tustin
1915		George L. Brooks* W.F. Delaney Business Manager	R.W. Hemming (Treasurer Ray Travis)	Frank Tustin
1916			W.F. Delaney	Harold Forrest
1917			W.F. Delaney	George C. Cooper
1918			R.W. Hemming	J.F. Little
1919			H.L. O'Neill	J.F. Little
1920–1925	*no records found*			
1926			R.R. Wilbourne	F.K. McGovern
1927			R.R. Wilbourne	J.V. McDonald
1928			R.R. Wilbourne	H.L. O'Neill
1929	*no records found*			
1930			H.L. O'Neill	H.B. Stallcop
1931			H.B. Stallcop	

* During the Reid-Murphy split, from 1908 to 1913, #77 was not part of the IBEW IO and what could be found of #77's records do not confirm its officers.

Year	Business Manager	President	Financial Secretary	Record Secretary
1932	Frank Tustin		Frank Tustin	Chris Christianson
1933	Frank Tustin		Frank Tustin	C.G. Payne
1934	Frank Tustin		Frank Tustin	C.G. Payne
1935	Frank Tustin		Frank Tustin	O.M. Anderson
1936–1937	George Mulkey		Frank Tustin	O.M. Anderson
1938	George Mulkey		Frank Tustin	Tom Arnold
1939	A.E. Martin		Frank Tustin	Tom Arnold
1940	A.E. Martin, H. Mullaney		Frank Tustin	Tom Arnold
1941–1942	A.E. Martin		Frank Tustin	Tom Arnold
1943–1945	A.E. Martin		Frank Tustin	G.S. Winter

Dam operator, Cabinet Gorge Dam, Idaho, early 1950s. Avista Archive

Local #77 Officers 1946–2013

Year	Business Manager	President	Vice President	Financial Secretary	Record Secretary	Treasurer
1946	Earl F. Wyatt			Frank Tustin	H.S. Silvernale	
1947	Earl F. Wyatt			Frank Tustin	G.S. Hobbs	
1948	Henry W. Newcomb			Ray Darling	G.S. Hobbs	
1949	C.N. Kunz	H.S. Silvernale (unconfirmed)		Ray Darling	G.S. Hobbs	
1950	Lloyd C. Smith	H.S. Silvernale (unconfirmed)		Lloyd C. Smith	K.R. Nathan	
1951	Lloyd C. Smith	H.S. Silvernale (unconfirmed)		Lloyd C. Smith	J.F. Flynn	
1952	Lloyd C. Smith	H.S. Silvernale (unconfirmed)		Lloyd C. Smith	J.F. Flynn	
1953	Lloyd C. Smith	H.S. Silvernale (unconfirmed)		Lloyd C. Smith	J.F. Flynn	J.H. Davis
1954–1959	Lloyd C. Smith	H.S. Silvernale (unconfirmed)		Lloyd C. Smith	J.F. Flynn	
1960–1970	Arthur B. Kenny	H.S. Silvernale (unconfirmed)		Arthur B. Kenny	Stan Bowen	
1971–1973	John L. Starcevich			John L. Starcevich	Stan Bowen	
1974	J Starcevich	Art Dakers, Jr.	Susan M. Zehnder	John L. Starcevich	Stan Bowen	E.M. Larson
1975–1976	John L. Starcevich			John L. Starcevich	Stan Bowen	
1976	John L. Starcevich			John L. Starcevich	Stan Bowen	
1977	Warren Adkins			Warren Adkins	Stan Bowen	
1978	Richard N. Rogers (appointed when Adkins resigned)	S.N. Hadley	James Valentine	Richard N. Rogers	Stan Bowen	E.M. Larson
1979	Richard N. Rogers	S.N. Hadley/ Byron G. Hood	James Valentine	Richard N. Rogers	Stan Bowen/ C.J. Shaffer	

Year	Business Manager	President	Vice President	Financial Secretary	Record Secretary	Treasurer
1980	Richard N. Rogers	Byron G. Hood	James Valentine	Richard N. Rogers	C.J. Shaffer	G. (Gale) F. Wirth
1981	Charles P. Silvernale (appointment)	Byron G. Hood	James Valentine	Richard N. Rogers	C.J. Shaffer	G. (Gale) F. Wirth
1982–1984	Charles P. Silvernale	Byron G. Hood	James Valentine	Charles P. Silvernale	C.J. Shaffer	G. (Gale) F. Wirth
1985	Charles P. Silvernale	James Valentine/Lou Walter	Lou Walter/ Richard Becker		C.J. Shaffer	G. (Gale) F. Wirth
1986	Charles P. Silvernale	Lou Walter	Richard Becker		C.J. Shaffer	G. (Gale) F. Wirth
1987	Charles P. Silvernale/Joe E. Murphy	Lou Walter/ Richard Becker	Richard Becker/ Richard Vaughn	*Fin Sec. & Bus. Mgr. position combined 1987*	Reuben Gable	Alan Sussex
1988	Joe E. Murphy	Richard Becker	Richard Vaughn	*same as Bus. Mgr.*	Reuben Gable	Alan Sussex
1989	Joe E. Murphy	Richard Becker	Richard Vaughn	*same as Bus. Mgr.*	Reuben Gable	Alan Sussex/ Danny L. Nelson
1990	Joe E. Murphy/Ray Warren	Richard Becker/ Richard Vaughn	Richard Vaughn/ Thomas Hayes	*same as Bus. Mgr.*	Reuben Gable	Danny L. Nelson
1991	Ray Warren	Richard Vaughn	Thomas Hayes	*same as Bus. Mgr.*	Reuben Gable	Marilyn Davis
1992	Ray Warren	Richard Vaughn	Thomas Hayes	*same as Bus. Mgr.*	Reuben Gable	Marilyn Davis
1993	Ray Warren	Richard Vaughn	Marilyn Davis	*same as Bus. Mgr.*	Reuben Gable	Sandra Polk
1994	Ray Warren/Don Guillot	Richard Vaughn	Marilyn Davis	*same as Bus. Mgr.*	Reuben Gable	Sandra Polk
				same as Bus. Mgr.		
1995	John Horrocks	Richard Vaughn	Marilyn Davis	*same as Bus. Mgr.*	Reuben Gable	Sandra Polk

Year	Business Manager	President	Vice President	Financial Secretary	Record Secretary	Treasurer
1996	John Horrocks	Richard Vaughn	Marilyn Davis-Westlund	*same as Bus. Mgr.*	Reuben Gable	Sandra Polk
1997	John Horrocks	Richard Vaughn	Marilyn Davis-Westlund/ Reuben Gable	*same as Bus. Mgr.*	Lou Walter	Sandra Polk
1998	John Horrocks/ Dave Timothy (appointed)	Richard Vaughn	Reuben Gable/Lou Walter	*same as Bus. Mgr.*	Lou Walter/ Sherman Williams Jr.	Sandra Polk
1999	Dave Timothy	Richard Vaughn	Lou Walter	*same as Bus. Mgr.*	Sherman Williams Jr.	Sandra Polk
2000	Dave Timothy	Richard Vaughn	Lou Walter	*same as Bus. Mgr.*	Sherman Williams Jr.	Sandra Polk
2001	Dave Timothy	Richard Vaughn	Lou Walter	*same as Bus. Mgr.*	Sherman Williams Jr.	Sandra Polk
2002	Dave Timothy	Richard Vaughn	Lou Walter	*same as Bus. Mgr.*	Sherman Williams Jr. /Sylvia A. Hanson	Sandra Polk
2003	Dave Timothy	Richard Vaughn	Lou Walter	*same as Bus. Mgr.*	Sylvia A. Hanson	Sandra Polk/ Joe Simpson
2004	Dave Timothy (resigned)/Don Guillot	Richard Vaughn/ Rick Johnson	Lou Walter /David Humlicek	*same as Bus. Mgr.*	Sylvia A. Hanson	Joe Simpson/ Saundra Selle
2005	Don Guillot	Rick Johnson	David Humlicek	*same as Bus. Mgr.*	Silvia A. Hanson/ David Wheeler	Saundra Selle
2006	Don Guillot	Rick Johnson	David Humlicek	*same as Bus. Mgr.*	David Wheeler	
2007–2009	Don Guillot	Rick Johnson	David Humlicek	*same as Bus. Mgr.*	David Wheeler	Lynne Moore
2008	Don Guillot	Rick Johnson	David Humlicek	*same as Bus. Mgr.*	David Wheeler	Lynne Moore
2009	Don Guillot	Rick Johnson	David Humlicek	*same as Bus. Mgr.*	David Wheeler	Lynne Moore
2010	Don Guillot	Rick Johnson		*same as Bus. Mgr.*	David Wheeler	

Year	Business Manager	President	Vice President	Financial Secretary	Record Secretary	Treasurer
2011	Don Guillot	Rick Johnson	Walter Aho	*same as Bus. Mgr.*	David Wheeler	Lynne Moore
2012	Don Guillot	Rick Johnson	Walter Aho	*same as Bus. Mgr.*	David Wheeler	Lynne Moore
2013	Don Guillot/Lou Walter	Rick Johnson	Walter Aho	*same as Bus. Mgr.*	David Wheeler	Lynne Moore

Laws Governing Collective Bargaining

The multitude of federal, state, and other administrative law that applies to #77's contracts and collective bargaining processes is both confusing and daunting.

This article is an attempt to outline as much of this law as possible. One of the webpages of the Washington State Public Employment Relations Commission (PERC) speaks to the difficulty in parsing out these laws on a page that compares various applicable state and federal statutes.

> The state of Washington has more collective bargaining statutes than any other state in the nation, and this compilation is for the convenience of Commission and staff members, as well as agency clientele who have occasion to work their way through that maze.
> Public Employment Relations Commission (PERC) website, "Statutory Comparison"

Here are a few definitions and terms that relate to collective bargaining.

Collective Bargaining

> Collective bargaining consists of negotiations between an employer and a group of employees so as to determine the conditions of employment. The result of collective bargaining procedures is a collective agreement. Employees are often represented in bargaining by a union or other labor organization. Collective bargaining is governed by federal and state statutory laws, administrative agency regulations, and judicial decisions. In areas where federal and state law overlap, state laws are preempted.
> Cornell Law School website, "Collective Bargaining and Labor Arbitration: An Overview"

Master Contract

A master contract is a contract negotiated by several affiliated unions who together negotiate as one party, representing all union workers, to address uniform/general terms of the contract. A master contract may cover all the union workers of one employer, and/or at one worksite. A master contract may be between one union and several similar employers, and is also called a multi-employer contract.

A master contract usually allows for specific (additional) terms to be negotiated by individual unions.

One example of a master contract is the Hanford Atomic Metal Trades Council (HAMTC) master contract. Local #77 is a signatory to this master contract. HAMTC is composed of fifteen different unions. The main body of this agreement (including benefits) is negotiated jointly. Appendix A has specific wages and working conditions for various affiliates, such as #77, which are negotiated separately.

Note that a number of Central Washinton PUDs have a "coordinated bargaining agreement." This differs from a master contract because it is an agreement between several employers and one union local (#77), not a trade council of affiliated craft unions and locals.

Labor management relations and negotiations are sometimes categorized as interest-based vs. positional bargaining.

Interest-based Bargaining

In this style of negotiation both sides present and negotiate from their interests. They then negotiate to find solutions that meet both sides' interests. Interest-based bargaining is also referred to as "getting to yes," or "collaborative" bargaining.An example of this style would be negotiating a change in work rules to increase profits without harming (or with improvements to) workers' wages, working conditions, and/or benefits.

Positional Bargaining

In this style of negotiation one or both sides come to the table with specific contract proposals.
An example of this style would be labor requesting a specific rate of pay, or management requesting a specific cut in hours.

Terms that can be collectively bargained are categorized as permissive subjects or mandatory subjects.

Permissive subjects

Permissive subjects are subjects that can be brought into negotiations only if both sides want to talk about them. Permissive subjects can be included in a contract, but are not required to be addressed.

Mandatory Subjects

Mandatory subjects are defined by applicable law. Such subjects can be brought into negotiations by either side. If they are brought forward the other side must negotiate them "in good faith."

A number of terms are used in relation to union membership at a worksite or under an employer. With passage of the Taft-Hartley Act in 1947, states were allowed to prohibit private sector "union shops."

Right-to-work-for-less, or right-to-work Laws

State laws, passed in twenty-five states to date, that remove any union membership requirement. Workers who are not in a union still receive the benefits of any union contract they may fall under. Idaho has a right-to-work-for-less law, described below.

Union Shop

A workplace where a worker must be a union member as a condition of employment at a time related to date of employment.

Closed Shop

A workplace where a worker must be a union member to be hired. Closed shops are illegal except where they existed before the Taft-Hartley Act and where the union has a hiring hall, which cannot operate exclusively for the union's members.

Open Shop

A workplace with a union contract, where a worker, although covered by this contract, may or may not be member of the union.

Agency Shop

A workplace where a worker within a bargaining unit can "opt out" by paying an "agency fee," an amount less than full union dues. (Such employees are *not* exempt from the contract.)

Here are the applicable laws and administrative bodies.

Applicable Federal Law

The Wagner Act/National Labor Relations Act (NLRA), passed in 1935.
The Taft-Hartley Act, passed in 1947.
The Labor Management Reporting and Disclosure (Landrum-Griffin) Act, passed in 1959.

These laws apply to most private non-agricultural employees and employers engaged in some aspect of interstate commerce. State workers are *not* covered by these federal laws because of constitutional concerns about state sovereignty.

Federal workers' collective bargaining rights fall under a variety of collective bargaining laws, not covered here.

The National Labor Relations Act (NLRA) established the National Labor Relations Board (NLRB) to hear disputes between employers and employees arising under the act and to determine which labor organization will represent a unit of employees. The act also establishes a General Council to independently investigate and prosecute cases against violators of the act before the NLRB.
Cornell Law School, Legal Information Institute (LII) website, "Labor Law: An Overview"

The following categories of #77 employers generally fall under these federal collective bargaining laws:
Rural Electric Associations (REAs).
"Investor-owned" utilities, also known as privately owned utilities.
Private contractors such as signatories to the Four Local Agreement. ("NECA-Four Local Agreement" refers to the agreement between the Northwest Line Constructors Chapter of the National Electrical Contractors Association (NECA) and four regional IBEW linemen's or construction locals, #77, #125, #483, and #659.)

Applicable Washington State Law
The Public Employment Relations Commission (PERC), which was created in 1975, covers all public employers and employees in the State of Washington with the exception of the Washington State Ferries System. Municipal, county, PUD, and Washington State employees who are members of #77 generally fall under PERC's jurisdiction. There are a variety of state laws administered by PERC, including the Public Employees' Collective Bargaining Act (RCW 41.56.900).

The Personnel System Reform Act of 2002 (RCW 41.80) brought state workers under PERC, and includes master agreements that apply to all agencies with employees who are in bargaining units represented by the same union.

Public Utility Districts (PUDs) have had an especially convoluted history of applicable collective bargaining law. In 1963, the Collective Bargaining Authorized for Employees (RCW 54.04.170) granted PUD workers the right "to enter into collective bargaining relations with their employers with all the rights and privileges incident thereto as are accorded to similar employees in private industry." Passed concurrently, Collective Bargaining Authorized for Districts (RCW 54.04.180) allows PUDS to "enter into collective bargaining relations with its employees in the same manner that a private employer might do and may agree to be bound by the result of such collective bargaining."

Applicable Idaho State Law
In 1985 Idaho passed a right-to-work-for-less law, Title 44, Chapter 20 Right to Work (see definition of "right-to-work," above). This law *does* allow local jurisdictions to adopt a collective bargaining ordinance. A union can also negotiate a recognition agreement with a specific local governing agency. Once such an agreement is signed, the union and the agency can collectively bargain under that agreement. Idaho's state employees can be represented by unions, but cannot collectively bargain.

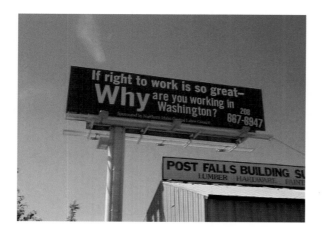

Left: Idaho passed right-to-work-for-less legislation in 1985. In response the North Idaho Labor Central Council designed and paid for this billboard. It faced drivers as they returned to Idaho from Spokane. Barbara Harris Collection.

Left from top: Longtime #77 business reps John Cunningham and John Trumble have between them negotiated almost every #77 contract. Photos circa 1998. Local #77 Archive.

Above and below: Lynne Moore on top of the SCL Whitehorse microwave dish, circa 1985. A journeyman iron worker, she was helicoptered in with her crew, then climbed, working over an approximately 1,800-foot drop. Moore is currently treasurer of #77. Gary Moore Collection.

IBEW Locals of the Pacific Northwest

Local Union	Location	Charter Date	Final Date	Final Activity	Notes
73	Spokane, WA	4/30/1894			At one time represented linemen. Currently represents Communications, Electrical, Manufacturing, Inside (wiremen).
76	Tacoma, WA	5/6/1894			Currently represents Inside (wiremen), Radio-Television Service, Sound Technicians.
77	Seattle, WA	8/28/1897			Currently represents Line Clearance, Tree Trimming, Maintenance, Outside (linemen), Telephone, Utility. #77, with #483, #125, and #659, is party to NECA-Four Local Agreement.
125	Portland, OR	10/11/1900			Currently represents Line Clearance, Tree Trimming, Outside (linemen), Telephone, Utility. #125, with #77, #483, and #659, is party to NECA-Four Local Agreement.
191	Everett, WA	7/8/1901			Currently represents Communications, Inside (wiremen), Maintenance and Operation, Radio-Television Broadcasting, Sound Technicians.
202	Seattle, WA	8/28/1901	8/20/1934		Amalgamated with a Local 9 but not #9 of Chicago.
217	Seattle, WA and/or Port Townsend	12/23/1901	12/1/1908	Defunct and/ or dissolved as of 4/21/1914	Held a first joint meeting with #77 and #202 just as #77 left the IBEW IO (as part of the Reid-Murphy split).
217	Port Angeles, WA	Unknown	Unknown		
334	Whatcom, WA	11/24/1902	4/1/1904	Defunct	

Local Union	Location	Charter Date	Final Date	Final Activity	Notes
458	Aberdeen, WA	1/1/1904	10/1/1959	Amalgamated with Local 76	
334	Bellingham, WA	5/1/1904	6/1/1905	Defunct	
314	Bellingham, WA	3/1/1906	12/1/1908	Defunct	
483	Tacoma, WA	3/8/1906			Currently represents Cable Television, Line Clearance, Tree Trimming, Outside (linemen), Utility. #483, with #77, #125, and #659, is party to NECA-Four Local Agreement.
523	Yakima, WA	10/1/1906	12/1/1908	Defunct	
556	Walla Walla, WA	5/23/1907	11/20/1940	Amalgamated with Local 73	
574	Bremerton, WA	7/25/1907			Currently in Silverdale, Washington, representing federal employees in the Puget Sound area.
580	Olympia, WA	8/13/1907	10/1/1958	Amalgamated with Local 76	
609	Spokane, WA	4/15/1908	7/11/1925	Amalgamated with Local 73	
685	Roslyn, WA	11/30/1910	12/1/1912	Defunct	
691	Spokane, WA	2/10/1911		Defunct	
48	Portland, OR	5/21/1913			Currently represents Communications, Electrical Manufacturing, Inside (wiremen), Radio-Television Broadcasting, Radio-Television Service, Sound Technicians. #659,with #77, #125, #483, and is party to NECA-Four Local Agreement.
30	Spokane, WA	6/4/1913		Amalgamated with Local 73	

Local Union	Location	Charter Date	Final Date	Final Activity	Notes
46	Seattle, WA	4/7/1914			Currently represents Alarm and Signal, Cable Splicers, Communications, Electric Signs, Electrical Manufacturing, Electronic Technicians, Inside (wiremen), Maintenance, Maintenance and Operation, Marine, Radio-Television Broadcasting, Radio-Television Service, Sound Technicians
13	Seattle, WA	Unknown	7/1/1914	Amalgamated with Local 46	
217	Port Angeles, WA	4/21/1914		Defunct	
359	Centralia, WA	12/28/1914		Defunct	
241	Seattle, WA	4/20/1917		Defunct	
721	Seattle, WA	7/24/1917		Defunct	
654	Tacoma, WA	7/30/1917	4/1/1923	Defunct	
441	Ellensburg, WA	1/14/1918		Defunct	
944	Seattle, WA	5/2/1919	12/29/1925	Changed to Local 77	
952	Seattle, WA	5/22/1919		Defunct	
459	Vancouver, WA	9/18/1919		Defunct	
1012	Ellensburg, WA	12/28/1919	11/1/1923	Defunct	
1086	Kittitas, WA	6/4/1920		Defunct	
1086	Tacoma, WA	10/25/1920	10/1/1980	Amalgamated with Local 1767	
1117	Seattle, WA	10/28/1920	October, 1921	Defunct	
1130	Port Angeles, WA	12/22/1920	June, 1922	Defunct	
970	Kelso-Longview, WA	4/3/1924	Unknown	Unknown	No longer on IBEW roster.

Local Union	Location	Charter Date	Final Date	Final Activity	Notes
843	Burlington, WA	11/25/1925	November, 1927	Amalgamated with Local 1032	
497	Wenatchee, WA	3/19/1926	4/1/1993	Amalgamated with Local 191	
828	Seattle, WA	3/12/1930	4/29/1931	Amalgamated with Local 46	
997	Port Angeles, WA	5/2/1934			Utility
745	Centralia + Chehalis, WA	12/26/1934	1/20/1936	Amalgamated with Local 77	
741	Seattle, WA	6/2/1936	7/22/1937	Amalgamated with Local 77	Referred to as radiomen's local in the April 1937 *Electrical Worker*.
882	Shelton, WA	6/2/1936	6/14/1960	Defunct	
803	Tacoma, WA	6/19/1936	5/1/1940	Amalgamated with Local 76	
659	Medford, OR	4/16/1937			Currently represents Cable Television, Communications, Electrical Manufacturing, Inside (wiremen), Line Clearance, Tree Trimming, Outside (linemen), Sound Technicians, Telephone, Utility.
562	Wenatchee, WA	1/6/1939	3/1/1957	Amalgamated with Local 1769	
112	Kennewick, WA	6/1/1947			Currently represents Communications, Inside Wiremen, Sound Technicians.
536	Spokane, WA	9/1/1948	2/1/1989	Amalgamated with Locals 1769 and 1856	
1665	Wenatchee, WA	9/16/1950	6/1/1977	Amalgamated with Local 497	

Local Union	Location	Charter Date	Final Date	Final Activity	Notes
1765	Seattle, WA	2/1/1952	8/1/1962	Amalgamated with Local 1769	
1767	Tacoma, WA	2/19/1952	11/1/1982	Amalgamated with Locals 1155 and 1769	
1769	Seattle, WA	3/1/1952			Currently represents railroad workers.
1782	Vancouver, WA	4/21/1952			Currently represents railroad workers.
984	Richland, WA	10/1/1956			Currently represents Atomic Research Service workers.
1958	Richland, WA	10/1/1956	12/31/1965	Amalgamated with Local 77	
1032	Bellingham, WA	Unknown	10/1/1961	Amalgamated with Local 191	
89	Seattle, WA	6/1/1967			Currently represents Cable Television, Electrical Manufacturing, Line Clearance, Tree Trimming, Outside, Telephone.
2329	Vancouver, WA	2/1/1971	5/5/1975	Defunct	

IBEW's District 9 leadership, probably at an IBEW International Convention. Local #77 Archive.

A few of those who helped the author along, clockwise from upper left: Current #77 Office Manager Kellie McGuire (circa 1998), Local #77 Archive. George and Maxine Bockman about the time he became a journeyman (1952), George Bockman Collection. Rick Johnson passing a sock line into the traveler of a 387-foot tower from a helicopter, for WICO Construction, 1980, Rick Johnson Collection. (A sock line is the rope line used to thread a traveler, or pulling wheel, for wire). Dick Shelley and his infamous red hat, Dick Shelley Collection. Don Guillot at work as #77 business manager, Local #77 Collection. Clyde Meredith just out of the Navy before hiring on with WWP, Clyde Meredith Collection.

Author's Note

FINDING balance between scholarship and the undocumentable truths of a community is always challenging. This book collects the stories and experiences of generations of IBEW #77 members, and attempts to provide historical context for them. Sources, be they personal accounts or corporate websites, have been cited wherever possible. When I couldn't find critical information I noted this.

It has been exhilarating and horrifying to see the role of chance in finding key original documents. Local #77, like most organizations and individuals, is busy being alive. So much of our history lurks in precarious stacks of cardboard and half-forgotten files, ripe for exploration.

Many people patiently tutored me on their work, their politics, their union, and their lives. It is an honor to be entrusted with their accounts. I suspect that in the heat of finishing this book I will overlook naming some of them. Consider those named here as examples of the support I received.

The Executive Board of #77 provided their total support from the minute they hired me.

Don Guillot gave me his vision for this book, offered constant encouragement, and told me terrible, terrible jokes. His candor, regardless of how it made anyone look, including himself, was heroic.

Rick Johnson prompted and cajoled me to delve further into just about every topic covered and made sure I had what I needed to do the work.

John Cunningham, a walking encyclopedia of #77's history and traditions, shared his commitment to progressive unionism and to #77's past, present, and future.

John Trumble laconically guided me through #77's legacy, leadership, and issues east of the Cascades with his expansive knowledge. May he never be far from the Kennewick #77 office, nor the pulse of the local.

Again and again, those I interviewed wanted to make sure I had a copy of *Proud Hands*, the first history of #77, written by Lou Walter. His roving mind was ahead of mine on so much of what is in this book, and what should be in the next volume about #77.

Suzanne Brunner supplied whatever I needed out of the Spokane area, from soup to early labor contracts. C.O. Smith, Rick Strait, Nichole Reedy, and Tom McMahon walked me through details and decades of employer relations, contract negotiations, and working conditions—then did it again when I got confused.

Jeff Cash, Rachel geBauer, Chris Martin, Joe Simpson, Steve Hendrickson, Bob Guenther, and John Logan answered every question I asked.

Some know enough to keep track of their (and our) history: Gary Moore, George Bockman, Buster Perry, Bill Lund, Clyde Meredith, Dick Shelley, Charlie Silvernale, Rick Luiten, Loren Noyes, Kathleen Merrigan, Heidi Durham, Lee Look, Nicole Grant, and Timm Kelly. Each of them illustrated their accounts of work, lingo, and equipment with items from their personal collections. Many others were kind enough to be interviewed, as the bibliography attests. Thank you.

Archivists are a special breed. Nicolette Bromberg, Elizabeth Russell, and Conor M. Casey patiently guided me through University of Washington Libraries' amazing Special Collections. The Northwest Museum of Arts & Culture/Eastern Washington State Historical Society's Jane Davey and Cheryl Gunselman at Washington State University Libraries' Manuscripts, Archives, and Special Collections were both extremely helpful.

Seattle City Light and Puget Sound Energy made timely donations to the University of Washington Library Special Collections that paid for creation of preliminary finding aids for their corporate collections. Shirley Wolf at Avista's Archive allowed me to use a digital copy of their collection.

Local #77 support staff always had a smile for me. Kellie McGuire, Shaunie Saelee, Diana Seals, Marcia Vogt, Nancy Greenup, Elaine Kinnear, and Mike Maloney proved what professionals they are. Linda Sansaver and Nancy Wangen found time to explain #77's dispatch system, even while dispatching.

Copy editor Angie Jabine remained stalwartly calm as reams of drafts came her way. The subtle and systematic accuracies in text and format are her work, the errors are mine.

My partner Doug Kilgore encouraged me, excused me, and kept my eyes on the prize with loving equanimity.

Finally, I want to thank every member of #77, past and present. It has been a true privilege to collect and document your history.

Ellie Belew
November 2013
Roslyn, Washington

Bringing Power to the People

Bibliography

Abbott, Arlene, NUTEC consultant. Interviewed by E. Belew, 2012–13.

"Agreement between the Northwest Line Constructors Chapter of the National Electrical Contractors Association (NECA) and International Brotherhood of Electrical Workers (AFL-CIO) Local Union No. 77 (Seattle, Washington), Local Union No. 125 Portland, Oregon), Local Union No. 483 (Tacoma, Washington), Local Union No. 659 (Central Point, Oregon)." Various versions 1946–current.

"Agreement between Spokane Electrical Contractors Assn., Portland Chapter National Electrical Contractors Assn., Seattle Electrical Contractors Assn., and International Brotherhood of Electrical Workers Local Unions No. B-77, Seattle, Wash., B-125 Portland, Oregon, B-659 Medford, Oregon, 483 Tacoma, Wash." 1946.

Arnesen, Eric. *Encyclopedia of U.S. Labor and Working-class History.* Vol. 1. 3 vols. New York: Routledge, 2007.

Bateman, Curtis. (IBEW Archivist). "LU 77 delegates." IBEW, 2012.

—. Interviewed by E. Belew, 2012.

Becker, Dick, retired #77 president and vice president. Interviewed by E. Belew, 2013.

Beder, Sharon. *History of the Business Roundtable 3.* 2012. http://www.herinst.org/ BusinessManagedDemocracy/government/national/history3.html (accessed 2013).

Billington, David P., Donald C. Jackson, and Martin V. Melosi. "The History of Large Federal Dams: Planning, Design, and Construction in the Era of Big Dams." Denver: U.S. Department of the Interior, Bureau of Reclamation, 2005.

Billington, Ken. *People, Politics, and Public Power.* Seattle, Washington: Washington Public Utilities Districts Association, 1988.

Blewett, Steve. *Building on a Century of Service.* Spokane, Washington: Washington Water Power Company, 1989.

Bluechel, Candice, Director, Northwest Utility Training and Education Center (NUTEC). Interviewed by E. Belew, 2013.

Bockman, George, retired #77 member. Interviewed by E. Belew, 2012.

Boston, Don, and Loren Manderscheid, current #77 members. Interviewed by E. Belew, 2012.

Brazier, Don. *History of the Washington Legislature 1854–1963.* Olympia, Washington: Washington State Senate, 2000.

Bremer, Richard. *Seattle in the 20th Century, Volume 1: Seattle 1900–1920: From Boomtown, Urban Turbulence, to Restoration.* Seattle, Washington: Charles Press, 1991.

Bremer, Richard. *Seattle in the 20th Century, Volume 2: Seattle 1921–1940: From Boom to Bust.* Seattle, Washington: Charles Press, 1992.

Bremer, Richard. *Seattle in the 20th Century, Volume 3: Seattle Transformed: World War II to Cold War.* Seattle, Washington: Charles Press, 1999.

Brier, Stephen, Project Director and Supervising Editor. *Who Built America? Volumes 1 and 2.* New York: Pantheon, 1992.

Brunner, Suzanne, current #77 business representative. Interviewed by E. Belew, 2012–13.

Cantelon, Philip L. "The Regulatory Dilemma of the Federal Power Commission, 1920–1977." *Federal History* (Society for History in Federal Government), No. 4 (January 2012): 65.

Cantrell, Carl, former director CIR/Bylaws and Appeals Department of IBEW. Interviewed by E. Belew, 2012.

Cash, Joe, current #77 business representative. Interviewed by E. Belew, 2012–13.

Chasan, Daniel Jack. *The Fall of the House of WPPSS.* Seattle, Washington: Sasquatch Publishing, 1985. Cited on Historylink website: http://www.historylink.org/index.cfm?DisplayPage=output.cfm&File_Id=5482 (accessed 2013).

Chelan County PUD. "History of Events Affecting Chelan County PUD". Chelan County PUD pamphlet, provided by Chelan County PUD to E. Belew, 2012.

Conover, Hank. Letter to J.F. Flynn, #77 recording secretary, April 1, 1955. IBEW Local #77 archives.

Cornell Law School. "Collective Bargaining and Labor Arbitration: An Overview." http://www.law.cornell.edu/wex/collective_bargaining (accessed 2013).

—. "Labor Law: An Overview." http://www.law.cornell.edu/wex/labor (accessed 2013).

Cunningham, John, #77 assistant business manager, business representative. Interviewed by E. Belew, 2012–13.

—. "One Local Craft Union and Gender Integration: 1974 to the Present." Research paper, 2010.

—. Personal archives.

De Rosa, Mike. "The Enron Debacle and Electric Power Deregulation." http://www.gp.org/articles/derosa_03_02_02.shtml (accessed 2013).

Dorpat, Paul and Genevieve McCoy. *Building Washington: A History of Washington State Public Works.* Seattle, Washington: Tartu Publications, 1998.

Duggins, Larry, former #77 executive board officer. Interviewed by E. Belew, 2012–2013.

Durham, Heidi. "Radical Women in Action—the Case of Seattle City Light." Radical Women pamphlet, 1975.

Easton, Steve, former #77 executive board officer. Interviewed by E. Belew, 2012–13.

Ennis, Bruce L. "Federal Power Commission Resolves Conflict between Priority and Preference in Favor of Private Power Producers." *26 Montana Law Review 246 (1964-1965)* http://heinonline.org/HOL/LandingPage?collection=&handle=hein.journals/montlr26&div=26&id=&page= (accessed 2013).

Erickson, Randy and Frank W. Marshall. "Chronological History of Development, Seattle Steam Company." 2004.

Federal Labor Relations Authority. "A Short History of the Statute." http://www.flra.gov/statute_history (accessed 2013).

Fichtner, Ray, retired #77 member. Interviewed by E. Belew, 2012.

Funding Universe. "Avista Corporation History." 2005. http://www.fundinguniverse.com/company-histories/avista-corporation-history/ (accessed 2013).

—. "Puget Sound Energy Inc. History." 2002. http://www.fundinguniverse.com/company-histories/puget-sound-energy-inc-history/ (accessed 2013).

Gable, Reuben, retired #77 treasurer. Interviewed by E. Belew, 2011.

geBauer, Rachel, current #77 business representative. Interviewed by E. Belew, 2012–13.

Grant, Nicole, executive director of the Certified Electrical Workers of Washington. Interviewed by E. Belew, 2013.

—. "Challenging Sexism at City Light: The Electrical Trades Trainee Program." Seattle Civil Rights & Labor History Project, http://depts.washington.edu/civilr. November 22, 2006. http://depts.washington.edu/civilr/citylight.htm (accessed 2013).

Greenspan, Alan. "The Reagan Legacy: Remarks, at the Ronald Reagan Library, Simi Valley, California, April 9, 2003." http://www.federalreserve.gov/boarddocs/speeches/2003/200304092/default.htm (accessed 2013)

Guenther, Bob. #77 governmental affairs advisor. Interviewed by E. Belew, 2012–13.

Guillot, Don, former #77 business manager. Interviewed by E. Belew, 2011–2013.

Haines, William Wister. Author and screenwriter of *Slim*, a novel and a Warner Bros. film production. "Slim" a lineman, is played by Henry Fonda. Released 1937.

Hanson, Mike, Avista utilities training coordinator, craft and technical-electric and/or generation operations, Interview by E. Belew, 2012.

Harris, Barbara, president of the Northern Idaho Central Labor Committee. Interviewed by E. Belew, 2013.

Harrison, John. "Bonneville Power Administration, History." Northwest Power and Conservation Council. 2008. http://www.nwcouncil.org/history/BPAHistory (accessed 2013).

Harvard Business School, Baker Library, Historical Collections, Lehman Brothers Collection. "The Washington Water Power Company." 2010. http://www.library.hbs.edu/hc/lehman/chrono. html?company=the_washington_water_power_company (accessed 2012).

Hendrickson, Steve, current #77 business representative. Interviewed by E. Belew, 2012–13.

Hirt, Paul W. *The Wired Northwest: The History of Electric Power, 1870s–1970s.* Lawrence, Kansas: University Press of Kansas, 2012.

Hughes, Thomas Parke. *Networks of Power: Electrification in Western Society, 1880–1930.* Baltimore, Maryland: John Hopkins University Press, 1993.

Humlicek, David, retired #77 member. Interviewed by E. Belew, 2013.

Hutchins, David, former #77 business representative. Interviewed by E. Belew, 2012.

International Brotherhood of Electrical Workers (IBEW). *Carrying the IBEW Dream into the 21st Century.* Washington, D.C.: IBEW, 2005.

—. "CIR/Bylaws & Appeals." http://www.ibew.org/IBEW/departments/CIR.htm (accessed 2013).

—. *The Electrical Worker*, 1893–2013. Archived at http://www.ibew.org/articles/menu/journal.htm (accessed 2013).

—. *History and Structure of the IBEW.* Washington, D.C.: IBEW, 1979.

—. "History of the IBEW: 1940–45" http://www.ibew.org/IBEW/history/1940_1945_3.htm (accessed 2013).

—. "Officers' Report." *Proceedings of the 34th Convention of the IBEW.* 1991.

—. "President Noonan's Remarks." *IBEW Convention Proceedings 1925.* IBEW, 1929.

—. "The Right Choice." http://www.ibew.org/index.asp (accessed 2013).

IBEW International archives. *IBEW LU 77.* 2011.

IBEW Local #73 History Committee, Harry Tuttle, Chair. "Mini History of IBEW Local B-73, 1896–1981." Spokane, Washington, circa 1981.

IBEW Local #77 archives. Local #77 offices, SeaTac, Washington.

IBEW Local #77. *With Proud Hands.* Seattle, Washington: IBEW #77, circa 1990.

Irvine, Rick, NUTEC instructor. Interviewed by E. Belew, 2013.

Isley, Bob, retired #77 business representative. Interviewed by E. Belew, 2012.

Johnson, Rick, IBEW #77 President. Interviewed by E. Belew, 2012–13.

Kelly, Timm, current #997 president and business agent, 2013.

Klinginsmith, Daryn, current #77 member. Interviewed by E. Belew, 2012.

Koreiva, Steve, current #77 member. Interviewed by E. Belew, 2012–13.

Kramer, Arthur. *Among the Livewires: 100 Years of Puget Power.* Edmonds, Washington: Creative Communications, 1986.

Kure, Stefanie. "IBEW: Lighting the Way from One Millennium to the Next." *Transmission & Distribution World*, August 2000.

Langdell, Gene, retired #77 business representative. Interviewed by E. Belew, 2012.

Leahy, Dan, public policy consultant and organizer of Progress Under Democracy. Interviewed by E. Belew, 2011–13.

Lee, Kai N. and Donna L. Klemka. *Electric Power and the Future of the Pacific Northwest.* Seattle, Washington: University of Washington Press, 1980.

Linder, Marc. *Wars of Attrition: Vietnam, the Business Roundtable, and the Decline of the Construction Unions.* Iowa City, Iowa: Fǎnpìhuà Press, 2000.

Locken, Art, retired #77 executive board member. Interviewed by E. Belew, 2012.

Logan, John, current #77 business representative. Interviewed by E. Belew, 2012.

Look, Lee, retired #77 member. Interviewed by E. Belew, 2012.

Luiten, Rick, current #77 assistant business manager. Interviewed by E. Belew, 2012–2013.

—. Personal archives.

Lund, Bill, retired #77 member. Interviewed by E. Belew, 2012.

Martin, Chris, current #77 organizer. Interviewed by E. Belew, 2012.

McMahon, Tom, current #77 business representative. Interviewed by E. Belew, 2012–13.

McNeil, John, retired #77 member. Interviewed by E. Belew, 2012.

McRoberts, Patrick. "Seattle General Strike, 1919." 1999. http://www.historylink.org/index.cfm?displaypage=output.cfm&file_id=861 (accessed 2013).

Meredith, Clyde, retired #77 member. Interviewed by E. Belew, 2012.

Merrigan, Kathleen, Heidi Durham, and Megan Cornish, retired #77 members. Interviewed by E. Belew, 2012.

Mitchell, Sidney Alexander. *S. Z. Mitchell and the Electrical Industry.* New York: Farrar, Straus & Cudahy, 1960.

Montana River Action. "The Failure of Montana Power Company." http://www.montanariveraction.org/sorry.saga.mpc.html (accessed 2013).

Moore, Gary, retired #77 member. Interviewed by E. Belew, 2012–13.

Mowrey, Mike, IBEW Ninth District International Vice President. Interviewed by E. Belew, 2012.

Murray, R. Emmett. *The Lexicon of Labor.* New York: The New Press, 1980.

Myers, Elaine and David Lee Myers. "Lessons from WPPSS: A $2.25 Billion Fiasco Illustrates the Drawbacks to 'Business as Usual' Approaches for Major Social Decisions." Originally published in *In Context, "Governance,"* Autumn 1984, p. 28 http://www.context.org/iclib/ic07/myers/ (accessed 2013).

National Rural Electric Cooperative Association (NRECA). "History of Electric Coops." December 2012. http://www.nreca.coop/about-electric-cooperatives/history-of-electric-co-ops/ (accessed 2013).

Northwest Power and Conservation Council. "Bonneville Power Administration, History." http://www.nwcouncil.org/history/BPAHistory.asp (accessed 2013).

—. "Rural Electrification." http://www.nwcouncil.org/history/RuralElectrification.asp (accessed 2012).

Northwest Washington Electrical Industry Joint Apprenticeship and Training Committee (JATC). *Northwest Washington Electrical Industry JATC.* http://www.nwejatc.org/ (accessed 2013).

Noyes, Loren, retired #77 member. Interviewed by E. Belew, 2012–13.

Nugent, Michael, retired IBEW IO archivist. Interviewed by E. Belew, 2012.

Nye, David E. *Electrifying America: Social Meanings of a New Technology, 1880–1940.* Cambridge, Massachusetts: The MIT Press, 1992.

Nygard, Bruce, retired #77 member. Interviewed by E. Belew, 2012.

Occupational Safety and Health Administration. "Timeline of OSHA's 40 Year History." http://www.osha.gov/osha40/timeline.html (accessed 2013).

Palladino, Grace. *Dreams of Dignity, Workers of Vision.* Washington, D.C: International Brotherhood of Electrical Workers, 1991.

Parkins, Terry, retired #77 member. Interviewed by E. Belew, 2012.

Perry, Buster (John), retired #77 member. Interviewed by E. Belew, 2012.

Platt, Harold L. *The Electric City: Energy and the Growth of the Chicago Area, 1880–1930.* Chicago: University of Chicago Press, 1991.

Polk, Sandra, former IBEW #77 treasurer. Interviewed by E. Belew, 2012.

Potelco, Inc. "Welcome to Our Site" http://www.potelco.net/aboutpotelco.html (accessed 2013).

Public Employment Relations Commission (PERC). "Statutory Comparison Composite Document." 2007. http://www.perc.wa.gov/statutorycomparison.asp (accessed 2013).

Public Power Council. "Public Power History." 2006. http://www.ppcpdx.org/in-ppHistory.html (accessed 2013).

Puget Sound Electricity. *Puget Sound Electric Journal.* March 1914.

Puget Sound Power and Light Company. "Puget Power and IBEW Local 77 Approve Contract." http://www.thefreelibrary.com/PUGET+POWER+AND+IBEW+LOCAL+77+APPROVE+CONTRACT-a014842770 (accessed 2013).

Reedy, Nichole, current #77 business representative. Interviewed by E. Belew, 2012–13.

Roberson, Amanda. "An Investigation of Post Avenue Steam Plants." Research paper, provided by Seattle Steam Company, June 2003.

Rogers, Dick. "George 'Lou' Brooks information presented to the IBEW International Offices." 1980.

Rudolf, Richard and Scott Ridley. *Power Struggle: The Hundred-year War over Electricity.* New York: Harper & Row, 1986.

Rural Electrification Administration. "A Brief History of the Rural Electric and Telephone Programs." U.S. Department of Agriculture, 1982. http://www.rurdev.usda.gov/rd/70th/legacy.html. (accessed 2013)

Seattle City Light. "A Brief History." http://www.seattle.gov/light/aboutus/history/ab5_brhs.htm (accessed 2013).

Seattle Steam Company. "Chronological History of Development." Seattle, Washington, 2009.

—. "Seattle Steam Background" pamphlet, Seattle, Washington, 2009.

Shaffer, Bill, retired #77 business representative. Interviewed by E. Belew, 2011.

Shelley, Dick, retired #77 member. Interviewed by E. Belew, 2012–13.

Silvernale, Charles, former #77 business manager. Interviewed by E. Belew, 2011–2013.

Simpson, Joe, current #77 business representative. Interviewed by E. Belew, 2012.

Smith, C.O., current #77 business representative. Interviewed by E. Belew, 2012–2013.

Smith, Michael, retired #77 member. Interviewed by E. Belew, 2012.

Smithsonian Institution, Powering a Generation of Change. "Emergence of Electrical Utilities in America." 2012. http://americanhistory.si.edu/powering/past/h1main.htm (accessed 2013).

Stone, Bill, former training director with NW Line Joint Apprenticeship and Training Committee (JATC). Interviewed by E. Belew, 2013.

Strait, Rick, current #77 construction business representative. Interviewed by E. Belew, 2012–13.

Tallman, Paul, retired #77 member. Interviewed by E. Belew, 2012.

Tennessee Valley Authority. "From the New Deal to a New Century." http://www.tva.com/abouttva/history.htm (accessed 2013).

The Advertiser (Adelaide, Australia). December 12, 1941.

Timothy, David, former #77 business manager. Interviewed by E. Belew, 2012.

Tollefson, Gene. *BPA and the Struggle for Power at Cost.* Bonneville Power Administration, 1987.

Trenti, Sonny, retired #77 member. Interviewed by E. Belew, 2012.

Trumble, John, current #77 business representative. Interviewed by E. Belew, 2012–2013.

U.S. Department of Energy. *Environmental Assessment: Expansion of the Volpentest Hazardous Materials Management and Emergency Response Training and Education Center, Hanford Site, Richland, Washington.* Richland, WA: U.S. Department of Energy, 2002.

U.S. Department of the Interior, Bureau of Reclamation, "Bonneville Power Act." http://www.usbr.gov/power/legislation/bonnevil.pdf (accessed 2013).

United Electrical Workers. "Who We Are: UE History in Brief." 2013. http://www.ueunion.org/uewho.html

—. "UE History in Brief." http://www.ranknfile-ue.org/uewho5.html (accessed 2013).

University of Denver. Howard Jenkins, Jr. website. http://www.law.du.edu/jenkins/Landrum.htm (accessed 2013).

University of Washington Special Collections. "Northwest Public Power Association, Acc. No. 214–11 Guide." 1989.

Voss, Jim, retired #77 member, Safety and Training Manager Electrical Safety Consultants International, Inc. Interviewed by E. Belew, 2013.

Walter, Lou, current #77 business manager. Interviewed by E. Belew, 2012–13.

Warren, Ray, former #77 business manager. Interviewed by E. Belew, 2012.

Washington Public Utility Districts Association. "History of PUDs in Washington." 2009. http://www.wpuda.org/pud-history.cfm (accessed 2013).

Washington State Department of Labor and Industries (L&I). "History of Apprenticeship." http://www.lni.wa.gov/TradesLicensing/Apprenticeship/About/History/default.asp (accessed 2013).

Washington State Labor Council, AFL-CIO. "State Employee Collective Bargaining." 2009. http://wslc.org/legis/stateemp.htm (accessed 2013).

—. "Joe Murphy (1940–2004)" WSLC Reports, November/December 2004. http://www.wslc.org/reports/Rep-0411.htm (accessed 2013).

Washington State Legislature. "Title 54 RCW Public Utility Districts." http://apps.leg.wa.gov/RCW/default.aspx?Cite=54# (accessed 2013).

Wikipedia. "American Federation of Labor." January 2013. http://en.wikipedia.org/wiki/American_Federation_of_Labor (accessed 2013).

—. "Energy Northwest." 2013. http://en.wikipedia.org/wiki/Energy_Northwest#History (accessed 2013).

—. "Grand Coulee Dam." 2013. http://en.wikipedia.org/wiki/Grand_Coulee_Dam#Background (accessed 2013).

—. "International Brotherhood of Electrical Workers- IBEW." 2013. http://en.wikipedia.org/wiki/IBEW (accessed January 2013).

—. "Industrial Workers of the World." 2013. http://en.wikipedia.org/wiki/Industrial_Workers_of_the_World (accessed 2013).

—. "Knights of Labor." 2013. http://en.wikipedia.org/wiki/Knights_of_Labor (accessed January 2013).

—. "Public Utility Holding Company Act of 1935." 2013. http://en.wikipedia.org/wiki/Public_Utility_Holding_Company_Act_of_1935 (accessed 2013).

—. "United Electrical, Radio and Machine Workers of America." 2013. http://en.wikipedia.org/wiki/United_Electrical, Radio_and_Machine_Workers_of_America.

Wilma, David W., Walt Crowley, and the HistoryLink Staff. *Power for the People, A History of Seattle City Light.* Seattle, Washington: History Ink/History Link in Association with the University of Washington Press, 2010.

Wilma, David. "Washington Public Power Supply System (WPPSS)." 2011 http://www.historylink.org/index.cfm?DisplayPage=output.cfm&File_Id=5482 (accessed 2013).

Wing, Robert C., Editor. *A Century of Service: The Puget Power Story.* Bellevue, Washington: Puget Sound Power & Light Company, 1987.

York, Anthony. "The Deregulation Debacle." January 30, 2001. http://www.salon.com/2001/01/30/deregulation_mess/ (accessed 2013).

Zinn, Howard. *A People's History of the United States: 1492–Present.* New York: HarperCollins, 2003.

Local #77 members walking on water as they inspect SCL dam platform of the Ross Dam, Washington, July 27, 1997. Gary Moore Collection.